C0-AYF-585

EMPLOYMENT POLICY

Employment Policy
The Crucial Years
1939–1955

Jim Tomlinson

CLARENDON PRESS · OXFORD
1987

Oxford University Press, Walton Street, Oxford OX2 6DP
Oxford New York Toronto
Delhi Bombay Calcutta Madras Karachi
Petaling Jaya Singapore Hong Kong Tokyo
Nairobi Dar es Salaam Cape Town
Melbourne Auckland

and associated companies in
Beirut Berlin Ibadan Nicosia

Oxford is a trade mark of Oxford University Press

Published in the United States
by Oxford University Press, New York

British Library Cataloguing in Publication Data
Tomlinson, Jim
Employment policy: the crucial years
1939–1955.
1. Labour policy—Great Britain—History
—20th century
I. Title
331.12'5'0941 HD8391
ISBN 0–19–828564–7

Library of Congress Cataloging in Publication Data
Tomlinson, Jim.
Employment policy.
Bibliography: p.
Includes index.
1. Great Britain—Economic policy—1918–1945.
2. Great Britain—Economic policy—1945–
3. Great Britain—Full employment policies. I. Title.
HC256.4.T586 1987 339.5'0941 87–7837
ISBN 0–19–828564–7

Set by Cambrian Typesetters, Frimley, Surrey
Printed and bound in
Great Britain by Biddles Ltd,
Guildford and Kings Lynn

Acknowledgements

Various parts of this book have benefited from the comments of Alan Booth, Bill Garside, Russell Jones, and Roger Middleton. Neil Rollings has not only offered helpful comments, but also guided me through some of the complexities of Public Record Office files.

Much of the research on which this book is based was undertaken while I was receiving a Personal Research Grant from the Economic and Social Research Council, no. E00242019.

Mrs Christine Newnham typed and retyped the manuscript with her customary speed and efficiency.

Contents

Introduction

The rise of national economic management in most advanced capitalist countries was a feature of the middle decades of the twentieth century. In Britain, management of the external account was a notable development of the interwar period, especially after the departure from the gold standard in 1931. Such external management, while a necessary condition for internal economic management, did not of course determine that such domestic management would be pursued, nor with what objectives.

Domestically, while governments historically were hardly indifferent to such issues as inflation and living standards, it was the rise of employment as a central policy concern that was criticial to the post-1945 emergence of the managed economy in its strong sense. This book is about the emergence of that policy concern and the changes in the actual conduct of policy that such a concern led to.

The book takes as its focus of interest the years 1939 to 1955. Before 1939 employment had already emerged as a major political issue, but the impact on policy had been slight. For the policy-making authorities in Britain, such political debate had been responded to largely by assertions of the market's capacity to resolve the employment problem. Such a view was certainly dominant in the 1920s, though by the mid-1930s it 'had been superseded by a more pessimistic philosophy which ascribed a much enlarged role to government and recognised the case for stabilisation.'[1] But while this was an essential transitional stage towards an employment policy, such a policy did not emerge before 1939. The favourable effects of some government policies, such as rearmament, on employment was not evidence of a movement to such an end as the primary purpose of policy.

By 1944, however, in the famous White Paper on Employment Policy[2] of that year, a clear turning-point was reached. Government not only essayed the possibility of pursuing employment policy, but also accepted a responsibility for maintaining employment, in a manner that interwar governments had most strenuously attempted to avoid. Thus, a central part of the book will look at the genesis of that White Paper and its precise significance. Clearly, it marked a substantial (and in itself important) shift in what may be called the rhetoric of public policy. There were compelling political reasons for such a change in the discourse of public policy.[3] But how far this reflected a fundamental reordering of the authorities' policy concerns is much less clear, and this is discussed in Chapter 3.

The shift in public policy rhetoric was signalled by a flood of literature on full employment. Most famous of these was Sir William Beveridge's *Full Employment in a Free Society*, also published in 1944. Chapter 4 compares this with the government White Paper, in order to bring out the specific character of the latter, as only one formulation of the possibilities of employment policy.

Chapter 5 looks at the implementation of the policy suggestions of the 1944 White Paper. How far were they a plausible framework for policy-makers to use in attempting to turn a general promise into a viable machinery of implementation?

Budgetary policy has always been central to employment policy in Britain. Chapter 6 examines the changes in budgetary policy of 1947, the year in which the use of the Budget emerged into the central role it was to maintain in economic management for the next forty years—and arguably beyond.[4] That budgetary policy matters, and that it changed substantially in 1947, is not disputed in this part of the book. Rather, the issue is how far this change marks a fundamental transition to the subordination of budgetary policy to the needs of employment which the notion of 'Keynesian revolution' in economic policy would seem to require.

The last years of the 1945–51 Labour government mark the beginnings of the acceptance that full employment was not just a much-desired goal, but something capable of attainment. What this growing acceptance meant for policy is explored in Chapter 7.

Labour's commitment to the priority of full employment is hardly to be doubted, though that, of course, does not prove that the commitment was the cause of the attainment of that end. The strength of the Conservatives' commitment to such a priority might be seen as more doubtful. They had presided over most of the years of mass unemployment in the 1920s and 1930s. The forcing of the issue in the war years owed much both to Labour's dominance of domestic policy in the Coalition, and to the wider, commonly non-Conservative, pressures to make sure the interwar experience should not be repeated.[5] Chapter 8 explores the Conservatives' approach to employment policy in their first postwar period of office. How far did the public rhetoric here conceal a continuing potential conflict over policy measures in the event of a return of mass unemployment?

The final chapter is an epilogue. If the first eight chapters chart the flowing tide of employment policy, in 1986 one must also record the ebb. While a systematic account of the decline of employment policy cannot be attempted here, a more limited exercise seems appropriate.[6] This involves a comparison between the 1944 White Paper, the high watermark in the belief in the possibility of defeating unemployment, and the 1985 White Paper, *Employment: The Challenge for the Nation*, which can stand as a summary of the current arguments for the incapacity of government to do much in this field. By means of this comparison, some of the basis for the loss of faith in the efficacy of employment policy may be suggested. Discussion of the possible means of recovery of such faith lies outside the aims of the book.[7]

Notes

[1] R. Middleton, *Towards the Managed Economy: Keynes, the Treasury, and Economic Policy in the 1930s* (London, 1985), 174.

[2] Cmd. 6527.

[3] P. Addison, *The Road to 1945* (London, 1977).

[4] A strong case can be made for seeing Mrs Thatcher's 'monetarism' as centrally a deflationary fiscal policy. See, for example, G. Thompson, *The Conservatives' Economic Policy* (London, 1986), Ch. 2.

[5] Addison, *The Road to 1945*.

[6] A more detailed account is given in J. Tomlinson, *British Macroeconomic Policy since 1940* (London, 1985), Chs. 1, 6.

[7] J. Tomlinson, *Monetarism: Is There an Alternative?* (Oxford, 1986).

1

Employment Policy before 1939

Employment as a central concern of economic policy is a twentieth-century phenomenon. Such centrality developed only slowly and was dependent upon a range of different conditions. At the most abstract level, such conditions include the very existence of labour markets, without which, of course, lack of employment could not arise. But for the purposes of this book, we must focus on a rather more limited time horizon and on more specific conditions.[1]

The growth of concern with employment is an effect of changes in the political agenda, the agenda of public policy-making. Change in this agenda, schematically, is the product of two further types of changes. On the one hand, this again changes with alterations in the structure of the political system—most notably the scope of the franchise, and the kinds of parties at work in the system. On the other hand, it changes with the kinds of political arguments that find a space within the political arena. The latter are of course themselves partly a function of the changes in political structure; for example, the ideology of socialism is conditioned partly by the existence of a mass electorate and political parties which seek to represent the working class. But the impact of political arguments also depends upon what we may, at this stage, broadly call institutional changes, which allow a relevance to certain arguments but make others seem beside the point. In the case of policy on employment, these institutional changes are particularly those within the economy.

This takes us on to the second set of conditions which made possible the dominance of employment in economic policy-making. These concern the growth of a national economy as an entity which could plausibly be subject to management. While an economy can be managed to a number of ends, the existence of the perceived possibility of national economic management makes employment policy a possibility.

Given these very general propositions, the approach of this chapter is to look at the development of employment policy before 1939 in the light of these two major kinds of changes: changes in political structure and ideas, and changes in economic structure. What is outlined is the slow, and by no means sure, emergence of unemployment both as a matter of public policy concern, and as a concern of specifically *economic* policy, something to be combated by economic measures.

The Emergence of Employment Policy

The beginnings of the modern concern with unemployment may be dated to the 1880s, after half a century in which 'unemployment as a serious theoretical and practical question was virtually ignored by English economic theorists and social reformers'.[2] It was the social reformers who mattered most, for it was largely through the concerns of social reformers that unemployment was established on the political agenda.

These concerns were above all poverty and its consequences—in the 1880s, crime, vagrancy, and prostitution. In later decades before the First World War those perceived consequences were expanded into racial degeneration and military unfitness. Unemployment, broadly speaking, could in this way take its place within the 'social' or 'Condition of England' question which so dominated domestic policy in the late Victorian and Edwardian periods. This gave scope for all kinds of policy stances on unemployment—repressive or reformative, individualist or 'social', charitable or state interventionist.[3] But all of these existed within a problematic of social administration, of relief and mitigation of the effects of unemployment, rather than economic prevention or cure.

To say why this was so entails the difficult task of explaining a negative. But it is illuminating in relation to the later development of employment policy to ask why, at this time, the widespread concern with unemployment did not lead to emphasis on public policy managing the economy in order to create employment.

In fact, such a proposal was made by the advocates of tariff reform. From the 1880s the case for protection ('Fair Trade') was argued partly on the grounds of its capacity to reduce unemployment. The same argument was used when the debate was revived under the aegis of Joseph Chamberlain in the early twentieth century. This advocacy failed both in the 1880s and the early 1900s. No simple reasons can be given for the failure, but two aspects stand out.[4]

First, tariff reform challenged the whole basis of Britain's policy relation to the world economy in the late nineteenth century. This was a policy stance, which consisted of an interconnected trio of politico-economic objectives: adherence to the gold standard, free capital flows, and free trade. Taken together, these three located Britain at the centre of an expanding world economy which not only generated economic benefits for most sections of British society but also fitted the dominant liberal political stance. This was a stance of internationalism, which held that increased economic interdependence also created political interdependence and reduced the chance of militarism and war.[5]

Against this predominant internationalist stance, tariff reformers could offer a 'Little Englandlism' which, while attractive to manufacturers hard hit by imports, could not offer a plausible alternative *political* economy. In particular, and this is the second crucial point, the attempt to pose protection as a solution to unemployment, and thereby as a means of capturing working-class support, not only ran up against the well-founded fears of the effects on working-class budgets of tariffs on food: in addition, as Semmel emphasizes, 'the leaders of the organised working class supported Free Trade not only as good economics, but because it was set in the political context of internationalism and peace'.[6]

These two points can be generalized. Tariff reform failed because it ran up against the 'brick wall' of Britain's complex and extensive intermeshing with the international economy. It proposed a form of national (or imperial) economic management which ran against the grain of Britain's posture of openness to the world economy. Equally, protection failed to capture the support of those who sought to represent working-class interests—those interests that might be expected to be most responsive to claims of giving priority to employment over other policy objectives (though there was, of course, opposition to the view that protection would be employment generating).

Retrospectively, the possibility of a successful direct political challenge to Britain's intermeshing with the world economy appears remote. Economically, the cyclical unemployment of organized workers in the export trades never prevailed long enough to open up a sustained questioning of Britain's trading relation to the world economy. Adherence to the gold standard was deflationary in the late nineteenth century, but with a major improvement in the terms of trade this price fall harmed few outside mortgagees in a dwindling agricultural sector.[7] While in principle free trade and gold standard adherence were separate policies, in Britain they were part and parcel of an overall economic and political orientation which few saw reason to challenge.

Those who would challenge it lacked intellectual resources. While economists were not very important in either public policy debate or policy-making,[8] it is notable that by and large they favoured free trade and the gold standard. While the latter was not at issue in this period, both policies could be defended by economists—on the same grounds—as favourable to a definite kind of liberal political economy. This is notably the case for both Pigou and Marshall. In his *Riddle of the Tariff* (1904), Pigou argued that under plausible assumptions an economic case for protection could be made—but that such a policy would require complex manipulations of the scope and scale of tariffs. 'Can we seriously suppose that a democratic Government, pressed upon all hands by interested suitors, bewildered by conflicting evidence, nervous of offending political adherents, would prove itself equal to that Herculean task?'[9] In this way Pigou followed Marshall, who also rejected tariffs on grounds that 'were mainly political in character'.[10]

If the tariff reformers lacked much in the way of intellectual resources to throw against the economic underpinnings of Pax Brittanica, they most notably also lacked a firm political base in the working class. We have already noted the dominance on the left of the same internationalist orientation that the tariff reformers faced in the liberal mainstream of British politics. The stance was linked to the general ideology of British socialism as it developed in the late nineteenth and early twentieth centuries. Although there was a 'national socialist' element in that socialism (e.g. Hyndman[11], the dominant element was a kind of left-wing liberalism. This combined a doughty defence of narrowly conceived working-class interests with a utopianism about the socialist future. What it lacked was a policy stance distinctive in objective or instruments (as opposed to rhetoric) from radicalism.[12]

This is clearly the case in the area of employment policy. At the level of rhetoric, the dominant view of British socialism was that unemployment was inevitable under capitalism; and the central thrust of socialist politics, especially at the parliamentary level, was for the recognition of the right to work.[13] Now as an ideological preamble to a substantive proposal, this was fine—but it lacked that follow-up. The major substantive proposal was for relief works, about which three points need to be stressed. First, they were by no means a new proposal. While economic theory might have little place for such endeavours, 'such theoretical constraints had often been thrust aside in times of commercial crisis'.[14] This was especially so after 1870, and from the 1880s this role for public works was recognized by a famous circular from the Local Government Board under Joseph Chamberlain.[15] Second, on all the evidence available at the time, such works could make only a marginal impact on employment. They were locally based and locally financed, and hence least affordable where most needed.[16] Their likely scale was therefore out of all proportion to the scale of the problem. Third, while the rhetoric was of the 'Right to Work', which many liberals found hard to swallow, the actual policy proposal of relief works showed how little the left had moved towards any notion of managing the economy as a means to securing employment. In domestic policy stance, as in international affairs, socialism at this time largely formed a kind of utopian extension of radical liberalism rather than a distinctive political and policy orientation.

Of course, before the First World War socialism in Britain was weak electorally, and so, whatever its policies, was most likely to function as a pressure group upon the dominant parties to take up issues that they otherwise might have played down.[17] But the point being made here is that socialism's electoral weakness was matched (though I am not suggesting *caused*) by an intellectual weakness, a lack of intellectual resources for discussing policy issues, which continued long after its electoral position was favourably transformed (see below).

As already noted, the dominant framework within which unemployment was discussed at this time was one of problems of poverty—of its prevention, and of the mitigation of its effects via policies of social administration. The section of the unemployed who fitted most readily into this definition of the problem were the casual workers—those in intermittent unemployment who swamped most urban labour markets. Most of the reforms addressed to the unemployment problem before the First World War saw the nub of that problem as dealing with the casuals—or preventing the skilled workers from slipping from a temporary interruption of work into the chronic state of casuality.

This is apparent in the Royal Commission on the Poor Laws, where casual labour is seen as the major cause of pauperism, itself the central concern of the Commission. In certain respects, the Minority Report does break new ground in respect of employment policy. Accepting the common hostility to local relief works as tainted by association with the Poor Law, the Report proposed counter-cyclical variation in work contracted by central government departments, with labour engaged in the normal way for standard wages. In addition, the Report raised the issue of financing such works from borrowings from underemployed capital during the depression, to be repaid in the boom. But while this was certainly an advance on most Labour Party thinking, it remained dominated by the traditional concern with poverty and its prevention, and by the notion of non-casual unemployment as essentially cyclical in character. This second position led to an underplaying of the scale of the challenge to existing institutional practices which solutions to unemployment might involve.[18]

Also important to unemployment policy in the Edwardian years was Beveridge's *Unemployment: A Problem of Industry*. Here unemployment was posed as an economic problem—in the sense of a problem of matching supply and demand in the labour market.[19] But again, the analysis was dominated by a concern for casuality and its effects, and was even further than the Minority Report from proposing a concept of a national economy to be managed in order to reduce unemployment.

The work of the Poor Law Minority (notably Beatrice Webb) and Beveridge provided the main intellectual underpinnings to the Liberal government's policies on unemployment. These involved three main strands:[20] first, labour exchanges to improve (including de-casualization) the labour market; second, national insurance for those in cyclical trades, to redistribute their incomes over the cycle, with a small element of government funding; finally, a policy of 'development works', combining the encouragement of works of 'national utility' (mainly roads and agriculture) with a possible counter-cyclical use.

Whatever else might be said of these proposals, they do represent a culmination of thirty years of fluctuating but generally growing concern about unemployment. Unemployment was on the agenda because of its implications for poverty; because of the attempts by sections of the left directly to mobilize the unemployed in a challenge to the existing order; and because the dominant political parties recognized the need to deal with issues that might be used to promote the electoral fortunes of the Labour Party. But the scope of policy to match this concern was narrow: narrow, because it took for granted the existing character of Britain's links to the international economy, and narrow because it did not face a political force able to challenge the existing institutional arrangements of the economy, or one with the intellectual resources to provide the arguments for such a challenge in the name of managing the economy.

The Impact of the First World War

The First World War brought substantial changes to both the economic environment and the politics of Britain, and together these changes refocused the employment issue. Initially, by far the most important change in the economy was the break with the gold standard. The exchange rate, which was pegged at around $4.76 from January 1916, was freed from its gold base in March 1919.

This was a direct political decision taken by Lloyd George's Cabinet, which was frightened by the prospect of unemployment, especially among demobilized servicemen. The decision was taken despite the firm commitment of the Bank of England and the Treasury to the ideal of a gold standard regime, demonstrated by the Report of the Cunliffe Committee.[21] As Morgan puts the point, 'External monetary considerations were for the present subordinate to domestic peace.'[22]

This episode not only demonstrated the political priority of unemployment at this time, but also signalled the first official recognition of the possibility of conflict between domestic employment and adherence to gold. Such a conflict, it is worth emphasizing, was never perceived to have arisen before the First World War. Once it was recognized, the way was open for the posing of employment policy as involving a break with the traditional orientation of Britain towards the world economy. In fact, however, the debates over unemployment policy in the 1920s were not dominated by such a perception of its possible implications. This is largely explained by the ambiguous effects of the war on the Labour Party.

Labour and the Economy

On the one hand, in an electoral sense, Labour's position was greatly advanced by the war. While its performance in the 1918 election yielded only some sixty seats, its vote had expanded to over 2 million and this was to continue to grow during the next few years, culminating of course in the first (minority) Labour government in 1924. For the leaders of the old parties, politics in the early postwar years was dominated by the 'threat from Labour'.[23]

Yet this electoral advance was accompanied by a profound conservatism in economic and financial policy, which greatly inhibited the Labour Party from providing plausible proposals for dealing with the mass unemployment that existed from 1921, after the postwar boom had collapsed.[24] Labour's leading financial spokesman (and chancellor of the Exchequer in 1924) was Philip Snowden. In 1920 he published a book entitled *Labour and National Finance*. The first sentence of this book summed up a whole world view: 'Sound finance is the basis of national and commercial prosperity.'[25] The book went on to berate the wartime government for borrowing to fight the war, rather than following the Gladstonian view that wars should be fought from current tax revenue. It called for a programme of public expenditure cuts and stressed the absolute priority of reducing the national debt.

This acceptance of financial orthodoxy included an acceptance of the need to return to gold. Thus, while Labour certainly regarded unemployment as a major issue in the early 1920s, its proposed remedies were markedly similar to those of the Liberals and Conservatives. These remedies focused on reviving the international economy, especially through an end to reparations and war debts and stabilization of trading relations with Germany and Russia.[26]

This profoundly conservative stance was accompanied by a commitment to counter-cyclical public works, financed by loans—though Snowden himself never seems to have accepted this form of finance. For the party leadership such works were seen as at best a temporary, stop-gap measure pending the reconstruction of the international economy. While the Independent Labour Party (ILP) developed a more ambitious programme of domestic reflation, it did not have a clear position on Britain's relation to the international economy and on how such domestic policies could be made compatible with the balance of payments constraint.[27]

Overall, therefore, we cannot see the early 1920s as a period when a growing Labour movement fought against prevailing economic orthodoxies. As Booth and Pack suggest, 'In spite of its reconstruction pledges, there is every reason to suspect that even a Labour Cabinet would have been swayed by orthodox financial pressures into defending the exchange rate at the cost of higher domestic unemployment.'[28] This stance of the Labour Party, it should be emphasized, was not just a consequence of its leaders' capture by the ideology of sound finance: it also reflected the lack of intellectual resources available to the party to draw upon in the area of economic policy. Even the best work in the Labour Party (mainly from the ILP) tended to be overly concerned with such issues as bank nationalization, whose relation to economic policy was extremely unclear. But beyond this, it also needs to be noted that financial orthodoxy was a deliberate strategic choice by the Labour leadership; for the overall strategy of the party at this time was one of demonstrating its 'fitness to govern'. This was clearly the case when Labour eventually came to power in 1924. In his period as prime minister, Macdonald's objective 'was to achieve such advances as were open to him within the framework of conventional politics and to convince anyone who might wish to that Socialism meant men of business pursuing the half measures constitutive of radical progress as England knew it'.[29]

Unemployment Policy in the 1920s

Lloyd George's tactical rejection of a return to gold in 1919 was replaced within just over a year by a deflationary policy which aimed at stemming domestic inflation and paving the way for a return to gold.[30] It is important to emphasize that there were practically no dissenters from this decision—Keynes, notably, was a 'dear money' man because of his fears of inflation, even though he was later the principal opponent of a return to gold. In addition, a return to gold was seen both as securing London's international financial position and as a means of constraining domestic policy action by 'irresponsible' politicians.[31] This last reason for advocacy of the gold standard is of particular importance in the context of this book, for, as discussed below, it was similar in many ways to the desire to pursue a balanced budget policy. In both policies, a central concern was to narrow the room for manœuvre by politicians—to limit the scope for deliberate political *management* of the economy. Thus, for a large part of the civil service, the City, and other ruling groups (including politicians themselves, it should be stressed[32]), there was an explicit, strategic, hostility to economic and financial management. Adherence to gold was thus a bulwark against such management. As Keynes put it, 'a chief object of stabilising the exchanges is to strap down ministers of finance'.[33] In fact, Keynes himself, while the chief advocate of a managed currency in the early 1920s, recognized the strength of this hostility to discretionary management, generated in part by the postwar inflation and financial instability. 'It is natural, after what we have experienced, that prudent people should desiderate a standard of value which is independent of finance ministers and state banks.'[34]

Movement back towards the gold standard was accompanied in the early 1920s by policies aimed at stabilizing the exchange rates of other countries and attempting to reduce the destabilizing impact of war debts and reparations. The aim was a return to pre-1913 international 'normalcy'. The perception that unemployment was a temporary phenomenon, caused by a normal trade cycle exacerbated by the disruption caused by the war, allowed space only for temporary palliatives. Hence the government followed the prewar tradition of public works programmes, under an Unemployment Grants Committee. This programme was small in scale and largely locally financed. The latter feature greatly limited its applicability, with unemployment generating the need for such works but also a loss of rate revenue for the local authorities who might have financed them.

The central feature of unemployment policy in this period was national insurance—that is, a policy aimed at the relief of unemployment, not its cure. In several stages this had been extended from its narrow base in 1911 to embrace most of the working class by 1920. Debates in the 1920s were dominated by the underlying fact that insurance in any strict sense was totally unable to cope with unemployment on the scale of this period. Hence all kinds of devices were adopted to try and keep the mass of unemployed workers out of the clutches of the Poor Law while maintaining the facade of a self-financing system. Enormous political and legislative time and energy were expended on the administration of unemployment relief. Much of the energy of the Labour Party in particular was expended on defending the right of workers to benefits and on liberalizing the terms and conditions of those benefits, rather than on policies for providing work. Equally important in the long run was the fact that, if the insurance scheme could not be self-financing, it must draw on the ordinary budget, and in this way come into conflict with the cause of sound finance. (This is, of course, precisely what occurred in 1931.)

Public Works and Public Finance

As already noted, relief works were deployed on a small scale in the early 1920s. Critics of policy favoured a much more extensive programme of such works. This was a consistent theme of Keynes (along with many other policy proposals) from 1924. It was also one favoured by many in the Labour Party. But such proposals found little favour among the leaders of either Labour or Conservative Party, or in any other quarters except the Liberal Party (returned to below).

Undoubtedly, such proposals were difficult to justify in terms of existing economic theory, which held that general involuntary unemployment was impossible in anything but the short run. In addition, in so far as such proposals were financed via public borrowing, they were held to 'crowd out' private investment and thus to have no overall positive effects on employment. Such a view did not lead to any clear-cut conclusions for economists, and indeed most economists favoured public works.[35] But this doctrine did find expression in the famous 'Treasury view', which had moved from growing coolness to a dogmatic hostility to public works by 1929.[36]

But to see the arguments over public works as fundamentally about economic theory is mistaken.[37] Much more important were what Middleton calls 'financial and administrative' constraints.

First, public works were opposed because of the implications they had for public borrowing. As noted above in relation to Snowden, the desire to reduce the national debt legacy of the First World War was very strong—not least at the Treasury. Such a debt was seen as requiring a taxation level that inhibited enterprise and investment, or threatened inflation if monetized. The strength of this feeling is illustrated by the breadth of support for a capital levy after the First World War, designed to pay off the debt.[38] Such a policy of debt reduction was, of course, entirely incompatible with any programme of public works financed by public borrowing.

Second, and even more fundamentally, the Treasury in particular saw the balanced budget rule as a defence against the tendency of politicians under democratic regimes to favour state expenditure but disfavour taxing the electorate to finance it.[39] Like an adherence to the gold standard, adherence to balanced budgets was seen by many civil servants as a way of preserving the long-term public interest against the short-sighted designs of politicians. Such a view undoubtedly gained a great deal of credence, as noted above, from the postwar inflations.

Third, opposition to budget deficits was based on the problem of 'confidence'. The Treasury in particular, taking private financing of investment as given, was able continuously to point out that, whatever the economic logic of public borrowing in a recession, if it undermined private investors' confidence then it might have adverse overall consequences. Thus the issue of 'crowding-out' in this view turned not on the existence or otherwise of 'idle balances', that is, unused finance for investment, but on the likely response of the private sector to perceived breaches of public sector financial probity.

These were the broad financial issues within which the conservatism of interwar policy on employment must be located. In addition, there were 'administrative' issues, which can best be illustrated by the arguments surrounding the 1929 general election.

Of all interwar elections, none was more focused on the unemployment issue than that of 1929. The Liberal Party made the running with its manifesto entitled *We Can Conquer Unemployment*. This in turn was very much based on the Liberal 'Yellow Book', *Britain's Industrial Future*, published in 1928. Recent analysis of these documents has suggested that 'The involvement of Keynes has given the Liberal proposals a lustre they do not merit.'[40] Such a view seems entirely justified. The proposals were large-scale relief works, on such tasks as road, bridge, and house-building. These works were very much a short run programme, to be implemented over two years. Theoretically, the programme differed little from the Treasury view. More important were the practical and administrative assumptions of the proposals.

First, and unlike Beatrice Webb's proposal of 1909 (above), the programme was aimed at directly employing the unemployed. This raised all sorts of difficulties about both the location of the works in relation to the unemployed, and the 'fit' between the labour required for such works and the characteristics of the unemployed.

Second, the envisaged speed of enactment of the proposals glossed over the means whereby such works would move from the planning stage to the beginning of actual building. Such movement raised issues of legal rights and administrative powers. While no doubt the government response to such proposals had a substantial element of self-serving in it, it is equally true that the Liberal proposals placed implausible emphasis on political will as the crucial determinant of what could be done in this area, especially in the time span suggested.[41]

In sum, the proponents of large-scale public works programmes were up against much more powerful forces than simply innate conservatism or the enthrallment of policy-makers to out-of-date ideas. To put the point another way, effective employment policy in this period would have required a much more radical, wide-ranging set of policies than that envisaged by any of the programmes on offer to the electorate in 1929. It would have required a challenge to the central *political* objectives of government policy, especially involving a re-orientation towards the domestic and away from the international economy, and an acceptance of a much larger role of the State in economy and society generally.[42]

The Labour Government, 1929–31

If the Liberal proposals of 1929 demonstrated the conservatism of even the most 'progressive' thinkers in their ranks, the Labour proposals demonstrated once again its combination of 'empty socialist rhetoric uneasily attached to a deeply orthodox economic analysis'.[43] Vague proposals for expansion were coupled with an attack on the Liberals' 'madcap finance', and the only definite promise was to set up a co-ordinating committee to oversee employment policy.[44]

Under the Labour government, the 'intractable million' unemployed, who had been the focus of attention in 1929, exploded to over 2 million. Unlike the Social Democrats in Sweden,[45] the Labour government had the misfortune to be elected *before* departure from the gold standard was forced by the financial collapse of 1931. It was thus constrained to pursue policies compatible with adherence to gold, in the absence of any viable alternative. This reinforced the already strong adherence to orthodox financial precepts regarding the Budget. By 1931 the budget deficit, especially that part resulting from the subsidy to the Unemployment Insurance Fund, had become the crucial measure of financial probity as perceived by international finance. Labour's inability to cope with this pressure led to the collapse of the government.

Between 1929 and 1931 Labour pursued policies not dissimilar to those of 1924. It expanded slightly the scale of relief works, and eased some of the conditions of the unemployment insurance scheme.[46] Beyond this, it pursued international policies aimed at reducing nationalistic pressures elsewhere in the world, pressures that were seen as dangerous both for international peace and economic stability. This latter stance meant that any policy of protection was ruled out along with the departure from gold. Labour remained committed to the traditional tenets of radical/liberal international policy, as it remained committed to sound finance at home.

Plainly, the Labour government was faced with unprecedented problems on the employment front. Any government faced with such an escalation of unemployment would have had grave difficulties. Nevertheless, Labour in 1929, while a minority government, was not constrained by the Liberals from pursuing a more radical policy. The election of 1929 had rejected the Conservative 'safety-first' policies and provided the parliamentary possibilities of a new programme on unemployment—which, while rapidly expanding in scale, was not of course a new problem in 1929. Once again, the failure of Labour to find the intellectual and political resources to break with its utopian/conservative stance must be emphasized. Snowden's view that 'socialism was a luxury that had to be financed out of revenue—like roads and other public utilities',[47] remained dominant. It is only one of the many ironies of this government that the objective that such financial orthodoxy was supposed to ensure—Britain's adherence to the gold standard—disappeared in September 1931 within a few weeks of the government's departure. The world did not come crashing around Britain's ears. But a New Era did open in economic policy and the possibility of economic management.

After the Gold Standard

In the interwar history of employment policy, the departure from gold in 1931 is the most significant single event. This is not because it immediately brought a revolution in economic policy: plainly, it did not. Rather, it is significant because of the space it allowed for some measure of deliberate economic management, and also for the 'intellectual space' it allowed for the flowering of all kinds of diverse proposals for such management.

The most striking aspect of the policies actually pursued in the 1930s aimed at reviving employment was their international orientation. While in the United States the 1930s brought a revolution in domestic policy, in Britain the decisive break with *laissez-faire* came in the sphere of international economic policy. The exchange rate, having floated down after 1931, was eventually held down deliberately to maintain competitiveness. The balance of payments constraint, and the dear money policy required to contain it, both disappeared in the 1930s. Cheap money at home and a managed exchange rate abroad replaced the 'automatic' gold standard. At the same time, general protection was instituted, bilateral trade bargaining pursued, and foreign investment restricted.[48] Thus, all the pillars of the Victorian international political economy were knocked down. Of course, none of this was part of a grand design. The departure from gold had been unwanted and unforeseen right up to the end. Protection had been awaiting an electoral opportunity since 1905, and the economic crisis delivered this. The other policies evolved on an *ad hoc* basis, mirroring those pursued in other countries. They amounted to an insulating wall around the economy, able to aid and abet a domestic recovery but not themselves the likely basis of such a trend.

Domestic policy after 1931 underwent no such revolution. In

broad terms the reasons are clear: none of the objections to expansionary fiscal policy discussed above disappeared. Expansionary monetary policy faced no such obstacles, and although cheap money was the order of the day, its power on its own to make substantial inroads into unemployment was small.

How far fiscal policy could have made much difference is a matter of dispute. Only recently have relatively sophisticated estimates of the expenditure multipliers been available, and these suggest that the required boost in government expenditure to make a substantial dent in the unemployment figures would have had to have been very large.[49]

Second there is the issue of confidence—how far could a deficit be pursued in the face of a threatened loss of faith in government policy by holders of sterling? Such a consideration was obviously crucial before departure from gold. Probably after 1931 this constraint was less tight, with a Conservative government in power and (from 1933) upward pressure on the exchange rate. But the anxious government pursuit of balanced budgets after 1931, via window-dressing, certainly suggests a continuing belief that sound finance had at least to be preserved in form, if less so in substance.[50]

Finally, there is the issue of the relevance to demand expansion for the highly structural and regional character of much of Britain's unemployment problem. This is partly an issue of timing. Such a demand expansion would have been most relevant when there were sharp cyclical downturns, especially that of 1929–32. One of the difficulties, as already suggested, is that what was required to such periods, but was not available, was a portfolio of projects already in an advanced state of preparation, to be brought rapidly into play. Such a programme, leaving aside other constraints, would have matched the character of unemployment in 1929–32; but for the rest of the interwar period its relevance must be more problematic.

In considering this issue, it is worth noting that, while demand management is a national policy, its first-round effects at least bear differentially on the economy, inasmuch as the initial effects depend on where the public works are located. Hence the favourable effects of rearmament were due partly to their major effects being in the most depressed regions.[59] Second, the effects of public works in particular regions would effect the rest of the economy on a scale depending upon the size of income leakages from those regions. Equally, the impact of this general reflationary movement on wages and hence inflation and export prices would depend on the degree of labour mobility—how easily people would move from the depressed to the more buoyant regions. On both these issues the evidence is sketchy and indecisive.[52]

All this remains necessarily counterfactual. Public works policies were not pursued in the 1930s. After 1931 there was a retreat from the fiscal conservatism of Snowden, notably in 1932 and 1933 with the continued growth of unemployment. Fiscal window-dressing was pursued, and the sinking fund suspended. But this retreat was very limited. Indeed, the enthusiastic pursuit of cheap money provided a further argument against any fiscal experiments—it might threaten cheap money.[53] Budgetary policy in the mid-1930s showed little effect from the permeation of Keynesian ideas. Up until the beginnings of rearmament in 1937, the emphasis on the balanced budget as central to the 'fiscal constitution'[54] remained, the maintenance of confidence was given priority, and experimentation was portrayed as the illusory easy answer.

Politically, the continuation of mass unemployment in the mid-1930s posed little threat to the governing party. Labour, but even more so the Liberals, suffered a crushing electoral defeat in 1931. Chamberlain, the chancellor from 1932, could present the slow and partial recovery from 1933 as the consequence of his 'sound' policies. The Labour Party had as yet little of a coherent alternative. Most of the political threat to the government arose from the issue of unemployment relief, reorganized in 1934/5 under the Unemployment Assistance Board.[55] Here Labour could plausibly present itself as the defender of the incomes of the unemployed; but it posed no political or intellectual threat to the policies that permitted this unemployment to continue.

Unintentionally, Labour also had another effect on employment policy. By its opposition to rearmament, it made it more difficult for such a policy to be pursued, and thus delayed what turned out to be the most effective method of reducing unemployment. In the same way that the party system in the 1920s constrained Labour from pursuing more radical employment policies, so in the 1930s the Conservatives were constrained from (unintentionally) reducing unemployment by rearmament in part by fear of pacifist feeling among the electorate.

Only from 1937 did the political situation change to allow a large-scale rearmament programme, which substantially reduced unemployment. But the implications of this should not be exaggerated. It is true that, once it became apparent that the rearmament programme was stimulating demand and employment, the Treasury manipulated the Budget to maximize this effect, and thus used the Budget to manage demand.[56] It is also true, of course, that this rearmament programme was financed largely by borrowing. But this did not lead to a 'conversion' of the Treasury to a belief in the use of the Budget, under ordinary peacetime conditions, to manage demand. 'Rather, it highlighted and reinforced what the Treasury had always maintained: that public expenditure could only be used, and would only be effective, as an employment measure in special circumstances, those dependent upon a favourable concatenation of political, economic and psychological factors.'[57]

As far as 'peaceful' public works are concerned, there was some movement towards acceptance of the *re-phasing* of such expenditure over the cycle in the downturn of 1937/8. But this implied little shift in the continued adherence to the balanced budget doctrine, or in the centrality of maintaining confidence. There is little evidence of a fundamental shift in policy stance on unemployment from the orthodoxy dominating the interwar period.[58]

Conclusions

Mass unemployment (as opposed to chronic underemployment) continuing over many years was a new phenomenon in the interwar period. The old problems of casual unemployment did not disappear, but they largely disappeared from view.[59] Unemployment became, for the first time, a problem of economic policy.[60] But the conditions for it to become the overriding determinant of policy did not exist. It did have such a priority briefly in 1919/20, in the face of Lloyd George's panic at the political situation; but for the next twenty years the undoubted public concern was not organized to challenge the political dominance of economic orthodoxy or the economic institutions that underlay such orthodoxy.

The forced departure from the gold standard in 1931 allowed 'a hundred flowers to bloom' in proposals for the new era of deliberate economic management.[61] But for the Treasury and ruling policians, the scope of this deliberate management was to be restricted largely to the external economy. Above all, the deliberate peacetime use of fiscal policy to manage demand had made little headway by the onset of war in 1939.

Notes

[1] For a much broader approach, see J. Garratty, *Unemployment in History: Economic Thought and Public Policy* (New York, 1978).

[2] J. Harris, *Unemployment and Politics: A Study in English Social Policy 1886–1914* (Oxford, 1972), 1.

[3] Ibid.

[4] B. Semmel, *Imperialism and Social Reform* (London, 1960).

[5] Classically the position of Cobden: see J. Morley, *The Life of Richard Cobden* (London, 1905), e.g. Ch. 29.

[6] Semmel, *Imperialism and Social Reform*, 107; A. W. Coats, 'Political Economy and the Tariff Reform Campaign of 1903', *Journal of Law and Economics*, 11 (1968), 214–5.

[7] Unlike in Argentina: see A. Ford, *The Gold Standard 1880–1914, Britain and Argentina*, (Oxford, 1962).

[8] Coats, 'Political Economy and the Tariff Reform Campaign of 1903', 181–229.

[9] A. C. Pigou, *Riddle of the Tariff* (London, 1904), 45.

[10] D. Winch, *Economics and Policy: A Historical Survey* (London, 1972), 68; Marshall's 'Memorandum on the Fiscal Policy of International Trade' is in J. M. Keynes (ed.), *Alfred Marshall's Official Papers* (London, 1926). Marshall argued that both US and German experience showed how tariffs corrupt the political process.

[11] Semmel, *Imperialism and Social Reform*.

[12] K. Middlemas, 'Edwardian Socialism', in D. Read (ed.), *Edwardian England* (London, 1982), 93–111. Ramsay Macdonald wrote of socialism at this time as follows: 'the stage which follows liberalism retains everything of permanent value that was in liberalism by virtue of its being the hereditary heir of liberalism'; cited in A. Briggs, 'The Political Scene', in S. Nowell-Smith (ed.), *Edwardian England 1900–1914* (Oxford, 1964), 66.

[13] K. D. Brown, *Labour and Unemployment* (Newton Abbot, 1971), 18–19.

[14] J. Harris, *Unemployment and Politics: A Study in English Social Policy, 1886–1914* (Oxford, 1972), 335.

[15] Ibid., 76.

[16] The Bill proposed that only in periods of exceptional distress should national schemes be launched, and even then they would be very much a departmental (rather than a Cabinet) concern. R. Skidelsky, *Politicians and the Slump* (Harmondsworth, 1970), 51–2.

[17] Brown, *Labour and Unemployment*, Conclusion.

[18] Thus Skidelsky, by taking it out of context, seems to overstate the significance of this policy (Skidelsky, *Politicians and the Slump*, 48–51). The Minority Report's proposals for 'The Regularisation of the National Demand for Labour' took up only three and a half pages out of 512, concerned primarily with the administrative issues arising from the decay of the Poor Law.

[19] J. Tomlinson, *Problems of British Economic Policy, 1870–1945* (London, 1981), 21–4.

[20] Harris, *Unemployment and Politics*, Ch. 6.

[21] J. Cunliffe, *1st Interim Report of the Committee on Currency and Foreign Exchange After the War*, Cd. 9182, 1918.

[22] K. O. Morgan, *Consensus and Disunity: The Lloyd George Coalition Government 1918–22* (Oxford 1979), 58.

[23] M. Cowling, *The Impact of Labour 1920–24* (Cambridge, 1971), Ch. 1.

[24] A. C. Pigou, *British Economic History 1914–1925* (London, 1947).

[25] P. Snowden, *Labour and National Finance* (London, 1920), 7.

[26] Morgan, *Consensus and Disunity*, 224.

[27] A. Booth and M. Pack, *Employment, Capital, and Economic Policy* (Oxford, 1985), 17–26; the ILP also focused on redistributive aspects of policy at the expense of the strictly macroeconomic; W. Garside, 'The Failure of the "Radical Alternative": Public Works, Deficit Finance and British Interwar Unemployment', *Journal of European Economic History*, 14 (1985).

[28] Booth and Pack, *Employment, Capital, and Economic Policy*, 11.

[29] Cowling, *The Impact of Labour*, 360.

[30] S. Howson, *Domestic Monetary Management in Britain 1919–38* (Cambridge, 1975), 11–23.

[31] Cunliffe, *1st Interim Report of the Committee on Currency and Foreign Exchanges after the War*, Cd. 9182. This is played down by S. Pollard (ed.), *The Gold Standard and Employment Policies Between the Wars* (London, 1970). Introduction.

[32] Booth and Pack, *Employment, Capital, and Economic Policy*, 28.

[33] J. M. Keynes, *A Tract on Monetary Reform* (Collected Writings, 4) (London, 1971), 136.

[34] Ibid., 135.

[35] T. W. Hutchison, *Economics and Economic Policy in Britain 1946–66* (New York, 1970), Appendix.

[36] Winch, *Economics and Policy*, 118–9.

[37] J. Tomlinson, 'Why Was There Never a Keynesian Revolution in Economic Policy?', *Economy and Society*, 10 (1981), 72–87; R. Middleton, 'The Treasury in the 1930s: Political and Administrative Constraints to Acceptance of the "New" Economics', *Oxford Economic Papers*, 34 (1982), 48–77.

[38] The extent of this support rested on the belief that a once-and-for-all *capital* tax would reduce the need for continuing high *income* tax, with the latter's perceived effects on incentives.

[39] R. Middleton, *Towards the Managed Economy: Keynes, the Treasury, and Economic Policy in the 1930s* (London, 1985), 87–91; see also Chapter 3 below.

[40] Booth and Pack, *Employment, Capital and Economic Policy*, 54.

[41] Tomlinson, *Problems of British Economic Policy 1870–1945*, 80–90.

[42] W. R. Garside, 'The Failure of the "Radical Alternative": Public Works, Deficit Finance, and British Interwar Unemployment', *Journal of European Economic History*, 14 (1985), 537–55.

[43] Booth and Pack, *Employment, Capital and Economic Policy*, 20.

[44] Skidelsky, *Politicians and the Slump*, 76–8.

[45] T. Skocpol and M. Weir, 'State Structures and the Possibilities for "Keynesian" Responses to the Great Depression in Sweden, Britain, and the United States', in P. Evans, D. Rueschmayer, and T. Skocpol (eds.), *Bringing the State Back In* (Cambridge, 1985), is an interesting attempt to compare Britain's and Sweden's response to the slump, though it plays down the importance of the departure from gold.

[44] Though this amelioration was heavily compromised by attacks on the benefits of certain groups, notably women: see J. Tomlinson, 'Women as Anomalies: The Anomalies Regulation Act of 1931, Its Background and Implications', *Public Administration*, 62 (1984), 423–37.

[47] Skidelsky, *Politicians and the Slump*, 95.

[48] H. W. Arndt, *Economic Consequences of the Nineteen Thirties* (New York, 1944), 94–134.

[49] T. J. Thomas, 'Aggregate Demand in the United Kingdom, 1918–1945', in R. Floud and D. McCloskey (eds.), *The Economic History of Britain since 1750*. II. *1860 to the 1970s* (Cambridge, 1981). 332–46; S. Glynn and P. Howells, 'Unemployment in the 1930s: The Keynesian "Solution" Reconsidered', *Australian Economic History Review*, 20 (1980), 28–45.

[50] Middleton, *Towards the Managed Economy*, 80–3.

[51] M. Thomas, 'Rearmament and Economic Recovery in the late 1930s', *Economic History Review*, 36 (1983), 552–79.

[52] For this debate see A. Booth and S. Glynn, 'Unemployment in Interwar Britain: A Case for Re-learning the Lessons of the 1930s?', *Economic History Review*, 36 (1983), 329–48; W. R. Garside and T. J. Hatton, 'Keynesian Policy and British Unemployment in the 1930s',

Economic History Review, 38 (1985), 83–8; A. Booth and S. Glynn, 'Building Counterfactual Pyramids', *Economic History Review*, 38 (1985), 89–94. On regional multipliers, see M. E. F. Jones, 'Regional Employment Multipliers, Regional Policy, and Structural Change in Interwar Britain', *Explorations in Economic History*, 22 (1985), 417–39.

[53] Middleton, *Towards the Managed Economy*, 114.

[54] J. Buchanan and R. Wagner, *Democracy in Deficit: The Political Legacy of Lord Keynes* (London, 1977).

[55] M. Cowling, *The Impact of Hitler* (Cambridge, 1975), 42–5.

[56] Middleton, *Towards the Managed Economy*, 120.

[57] Ibid., 121.

[58] Ibid., 165–71.

[59] R. C. Davison, *The Unemployed: Old Policies and New* (London, 1929), 178–81.

[60] Tomlinson, *Problems of British Economic Policy 1870–1945*, Ch. 4. Of course, there was still much 'social administrative' work on unemployment, e.g. Davison, *The Unemployed*, and Pilgrim Trust, *Men Without Work* (Cambridge, 1938). And, as argued above, much political attention fell on policies of amelioration of the social consequences of unemployment.

[61] For an assessment of some of these, see Booth and Pack, *Employment, Capital and Economic Policy*.

2

The Background to Employment Policy 1939–1955

This chapter provides a background to the detailed discussion of employment policy in the following chapters. The aim is not to give a comprehensive economic history of the period, but rather, to sketch the broad trends, highlighting those with special relevance to employment policy.

The Impact of War

Like the First World War, the Second saw a rapid expansion of the government role in the economy. In the First World War this had been very much a reluctant, piecemeal, process as government responded to new situations by grasping for new solutions. In the Second World War the process was much more rapid. However, it is not the case that the economy was transformed on to a 'total war' basis immediately in September 1939. Between the wars, low-key discussions of economic strategy in time of war had taken place, and lessons from the First World War had been learnt.[1] But, just as the military strategists accepted only reluctantly the impossibility of a 'war of limited liability', so only reluctantly did economic strategy completely give up any idea of business as usual.[2]

One of the lessons drawn from the First World War was the importance of minimizing inflation. Inflation was seen as a major source of social unrest in the earlier war. In addition, it was argued that inflation could not perform the necessary economic function of reducing the real incomes of the consumer, and hence making resources available for war purposes, because the strength of collective bargaining would mean that any price increases would be matched by wage increases. Hence the prime task of financial policy came to be to 'weaken the roots of inflation'.[3] From this strategy there followed a revolution in fiscal policy.[4]

Fiscal policy traditionally had been dominated by a concern to find financial resources to pay for government expenditure. Though, of course, the effects of fiscal policy on the economy generally and on employment in particular had been much discussed between the wars, the traditional form of fiscal calculation still dominated at the outbreak of war. The initial war budgets were based primarily on this concern: how to finance war expenditure without burdening posterity too greatly by excessive borrowing. But within the government, the increase in taxes that such a view entailed could also be seen in a different light—as a way of reducing the demand for real resources, and hence freeing those resources for war use, without inflation.

This approach to the budget was formalized and publicized in Keynes's article on 'How to Pay for the War', published originally in *The Times* in November 1939 and as a pamphlet in the following spring. Keynes's approach was to try and quantify the total resources available in the year, and to compare this with all the demands likely to be made on those resources. The clear implication of these calculations was that a drastic reduction of personal consumption would have to take place if the war were to be fought without inflation.[5] This approach was used in a full-blooded way from the Budget of 1941, commonly called the 'first Keynesian Budget'. What was new in 1941 was the 'universal acceptance of the Keynesian formulation and the existence of Keynesian arithmetic as one weighty element—but only one element—in the emergence of the decisive hunches'.[6]

It is important to emphasize the basis of this change in budgetary policy, because the change was to be a lasting one, and was (eventually) to dominate in the postwar decades. It was a change that revolutionized the budget calculation: this could now be integrated with the national accounts, to provide what can be sensibly called a macroeconomic budget. But the objectives of this calculative revolution fitted very much into the traditional objectives of Treasury policy—the minimizing of inflation. Hence the policy implications of this new arithmetic, heavy taxation, and compulsory saving to minimize borrowing did not run counter to the way Treasury policy would have likely gone anyway. Keynesian arithmetic provided a theoretical rationale and a quantification for policies that ran with rather than against the grain of the traditional Treasury stance on wartime finance.

The transformation of budgetary policy was coupled with a growth in direct planning of the economy which is the other striking change in economic policy in the war years. While budgetary policy in its new form attempted to provide the macroeconomic conditions appropriate for pursuing the war effort, the actual planning of that effort was done almost wholly in physical terms, and increasingly, as the war went on, in terms of manpower. But it would be wrong to see the wartime British economy as having a comprehensive planning apparatus. While, as with financial policy, experience of the First World War led to a greater degree of preparation, planning still had a cumulative but always partial character as in the earlier war.[7] Planning usually began on a relatively narrow front, but then extended backwards and outwards to embrace suppliers and competing users of inputs. It was especially important where shortages were most severe—initially, in British imports and manpower.

Planning and control of imports derived from a shortage of both shipping space and foreign exchange. By the end of 1940, practically all imports were purchased either directly by government ministries or under licence from the Board of Trade. Ultimately, control extended to 97 per cent of all imports, and nearly two-thirds were brought on government account. Even more striking was the scale of regulation of labour. From 1942, comprehensive manpower planning was the centrepiece of resource decisions in the economy, though always tempered by the desire to maintain labour's consent. Powers existed giving the Minister of Labour and National Service the right to control the registration, direction, and terms of engagement and dismissal of all workers. Manpower budgeting 'was, in fact, the only method the War Cabinet ever possessed of determining the balance of the whole war economy by a central and direct allocation of physical resources among the various sectors'.[8]

Attempts to allocate imports and labour soon translated into shortages of particular goods and services, which then led to those goods being rationed. Eventually major inputs into production came to be allocated centrally, while essential consumption goods were subject to strict rationing.

This growth of an enormous web of planning was co-ordinated from 1941 by the Lord President's Committee, which became in effect the Cabinet for home policy. The importance of this was that it marked the displacement of the Treasury from a hitherto dominant role in economic policy. Concern with finance was seen as at best secondary in the successful pursuit of the war effort.[9]

This extension of the government's role in the economy was accompanied by, and substantially conditional upon, a vast expansion of the role of economists in the government machine. Prior to 1939 professional economists were employed only in the agricultural ministries, though the Treasury was privy to discussions in the Economic Advisory Council, established by Macdonald in 1931, and its successor the Committee on Economic Information. At the start of war this led into the Stamp Survey and later into the Central Economic Information Service. Eventually, greatly expanded in size, there was created the Economic Section of the War Cabinet and the Central Statistical Office (CSO). The former body was during the war independent of the Treasury, and often dealt directly with the chairman of the Lord President's Committee. Much of the initiative on economic policy during the war years derived from the Economic Section[10]

The CSO was also very important in the new policy regime. Both the new fiscal policy and the planning of production required statistics on the economy to an unparalleled scale. The CSO played a vital role in this, particularly in regard to national income statistics for macroeconomic policy. Also of note was the role of economists in the Prime Minister's Statistical Section, under Lord Cherwell, which in addition to its scientific role had a considerable role particularly in the quantification and assessment of the broadly microeconomic aspects of the war effort.[11]

This accumulation of economic functions and personnel at the government level were both cause and consequence of a growth in ideologies of economic management. The departure from gold in 1931 had opened up a space for discussions of economic management in the 1930s, though the actual policy effects of this before the war were confined largely to altering Britain's role in the world economy, with little effort on domestic policy (see Chapter 1 above). The war opened the way for management to be applied domestically and on a scale unenvisaged in the years of peace. Of course, this application was to the peculiar problems of war, and in logic had no necessary implication for a peacetime economy. Unlike after the First World War, however, the success of the government role in the economy during war was now contrasted with the failures of the 1930s. In 1918 it was explicable, if unrealistic, to want to go back to 1913; in 1945, few wanted to go back to the 1930s. Thus there emerged by the end of the war 'the triumphant feeling that the economy had been successfully bent to the purpose of victory [which] led to the idea of bending it to another purpose, "reconstruction", in which the interests of all might play a part'.[12]

It is important to emphasize that the methods of national income management and physical planning used during the war owed little to the policy discussions of the 1930s. Nevertheless, they drew on a rationalistic and technocratic view of the possibilities of applying 'science' to the solution of social problems, which needs to be noted as an important condition of the postwar reconstruction.[13] Alongside this enhanced view of the capacities of rational intelligence to solve human problems was the growth of a drive for social reform and reconstruction, which is a striking feature from surprisingly early in the war years.

This drive for social reform had an undoubted political dimension: it represented the growing strength of the anti-Conservative forces which was to culminate in the Labour Party victory in the 1945 general election. Part of the drive for reform came from the generally enhanced role of Labour and the trade unions in the conduct of the war, and most particularly from the Labour members of the Cabinet, who dominated domestic policy-making in many areas. But this directly party-policial element should not be exaggerated. First, the most important wartime reformist plans derived not from the Labour Party but from Liberalism, in the form of Beveridge and Keynes. If one can reduce the social reforms of the 1940s to a simple formula, it would be a Liberal one—of reforming capitalism via enhanced state intervention but without threatening private property.[14]

More generally, the drive for social reform owed much to a cross-party consensus that modern 'total' war both exposed the scope for reform and imposed a political necessity to promise such reform, in order to motivate the populace who were all now, effectively, combatants. This view was apparent from the very outbreak of war, and though it occasionally irritated Churchill it came to be embraced by a substantial section of the Conservative Party.[15]

This linking of war and social reform had its particular

mechanisms of dissemination, embracing notably the BBC, the Army Bureau of Current Affairs, and the government information services. Together, these linked the shifts in popular opinion and those in 'official' opinion which together justify seeing the period from 1940 onwards as the beginnings of a new political consensus of which the election of Labour in 1945 was a rather delayed manifestation.[16]

Recently these wartime plans for social reform have been described as the offering of a 'New Jerusalem', totally unjustified by the weakness of the British economy, itself longstanding, but greatly exacerbated by the war.[17] Certainly, the general picture of the British economy by the end of the war is a dire one.

The British Economy in 1945

The problem was not one of a fall in total output. With the full employment of resources (unemployment fell to around 70,000 at its lowest point in 1943), and despite at peak the deployment of 4 million people into the armed forces, gross national product (GNP) expanded substantially during the war—from £5.2 billion in 1948 to £6.3 billion in the last full year of war, 1944.[18] Within this growth, consumption was squeezed—falling from £4.4 billion in 1948 to £3.9 billion in 1944. But this fall in consumption could not free sufficient resources for the war effort, and to a substantial extent Britain lived off its capital. Where gross investment (fixed capital and stocks) had been running at £500—£600 million per annum in the 1930s, at its low point it fell to well under £100 million in 1944, way below capital consumption of just under £400 million.[19] Overall, the domestic capital stock fell by approximately 10 per cent during the war.[20]

Even more pressing ws the balance of payments position. Here the war was sustainable only because of Lend-Lease from the United States and Canada. Merchandise exports fell to around 40 per cent of their prewar value while merchandise imports stayed at around the same value (they fell by around 40 per cent by volume). The current balance moved from an average deficit of around £30 million in the 1930s to £870 million by 1944 (excluding Lend-Lease supplies). Liquidation of foreign assets, especially in the United States, could not offset much of this current deficit, and itself reduced current revenue both during the war and after.[21] The situation was rendered much more serious by the abrupt and unexpected cessation of Lend-Lease immediately on the Japanese surrender in August 1945, which meant that, very quickly, the British economy had to not only restore but also surpass the prewar level of exports by perhaps 75 per cent. This reflected the loss of foreign assets, the loss of receipts from shipping, and the accumulation of debts in India and Egypt.[22]

The undoubted seriousness of this position cannot of itself justify the view that the attempt at fundamental social reform at the end of the war was an expensive luxury which Britain could not afford. Even if one accepts the highly problematic view that social reform retards economic growth, and that resources devoted to such reform represent a simple subtraction from those available for investment, the case against such reform is still not made. First, the extent to which reform did lead to the sacrifice of other economic objectives can be viewed only retrospectively, looking back at the period of the late 1940s, when most of the reform was implemented, and seeing what happened at this time to other economic indicators.[23] This is done below, in looking at the policies of the Labour government of 1945–51. Second, if, as many believed, the promise of social reform was necessary for morale in order to win the war, it is impossible to argue that such reform was an optional extra which could and should have been discarded. Whether or not such a view about the relationship between social reform and morale is appropriate, it was clearly widely shared. Hancock and Gowing aptly summarize this idea:

> There existed, so to speak, an implied contract between Government and people; the people refused none of the sacrifices that the Government demanded from them for the winning of the war; in return they expected that the Government should show imagination and seriousness in preparing for the restoration and improvement of the nation's well being when the war had been won. The plans for reconstruction were, therefore, a real part of the war effort.[24]

The Labour Government, 1945–51

The Labour government's period in office was marked by periodic economic crises, especially relating to the balance of payments position. Despite this, it can be argued that, by the end of the Labour government in 1951, the balance of payments problem was in an important sense resolved. In fact, current account balance was achieved by 1948, but it took the dollar appreciation of 1949 and the fall in import prices after the Korean war to approach some kind of equilibrium with the United States.

This success on the balance of payments front was all the more surprising because of the deterioration in the terms of trade, which left imports costing around 30 per cent more in terms of exports in 1951 compared with 1945.[25] The major expansion of exports (visible and invisible) was into the sterling area, while imports from the dollar area were cut sharply. Broadly speaking, the old interwar pattern of running a surplus with the sterling area to finance a deficit with North America was slowly restored, although the sterling area as a whole had difficulty earning dollars. Equally, the traditional British pattern of a visible trade deficit offset by an invisible surplus sufficient to finance substantial foreign investment (or debt repayment) was restored in 1950.[26]

While a new crisis blew up in 1951 because of the impact of the Korean war on world trade, the British economy by 1950 was broadly in a position to finance a full-employment level of imports in a way that seemed difficult to believe possible in 1945. Balance of payments crises did occur in the 1950s, but these were related largely to the sterling balances, that is, to the accumulations of sterling assets by India and Egypt during the war, and to a continuation of overseas military expenditure on an anachronistic imperial scale.[27] Trade uncompetitiveness was not a manifest feature of the British economy until the early 1960s.

The fact that the crucial change in trade was the reduction of imports from the dollar area also suggests a success for government policy. This was one area of policy where an objective was pursued systematically and consistently. It was also an area where economic planning was unambiguously successful.[28]

The rebuilding of Britain's trading position was greatly aided by both the damaged nature of the German and Japanese economies, and the fast growth of world trade. In the longer run it was this latter factor that was to sustain the boom conditions once the postwar restocking process was completed. Britain followed the United States in seeing the long-run benefits of such trade growth as best encouraged by an open international economy, rejecting the barriers to trade and capital flows which were perceived as having been both economically and politically damaging in the

1930s. However, this position 'in principle' was substantially breached for almost all the life time of the Labour government.

Britain's position on the structure of the postwar world economy was conditioned by a combination of this general ideological preference for *laissez-passer*, a very clear awareness of the difficulties that such an open economy would pose for Britain's balance of payments in the short run, and also the capacity of the United States to apply leverage over Britain's policy stance. On trade, it had been written into the Mutual Aid Agreement which followed Lend-Lease that Britain would end discriminatory and protective trade policies after the war—a reference particularly to the imperial preference arrangements of the 1930s. Equally, the US loan of 1945, which financed dollar imports once Lend-Lease had ceased, included a clause committing Britain to an end to restrictions on currency convertibility.

Despite periodic attempts to revive and enforce the free trade provisions of the Mutual Aid Agreement, Britain continued to discriminate against the dollar throughout the Labour government's period of office. The role of the sterling area as a trading bloc was enhanced rather than reduced, and a new role was played by Western Europe which was also built up as a trading bloc in this period.

On the question of currency convertibility, Britain was eventually persuaded to pursue this, in 1947. The result was a disastrous run on the pound which led to a resumption of controls after only six weeks. Convertibility was not restored again until 1955.

Thus, it cannot be said that British domestic policies during this period were effectively seriously inhibited by international commitments. In the negotiations over the postwar international institutions, Britain's fear was predominantly of a limiting of her capacity to resist a US slump, with all its consequences for employment.[29] This problem did not arise under the Labour government, though the small downturn in the US economy in 1949 kept alive the fears of the potential impact of events in the United States.[30]

Domestically, the compelling economic problems of the Labour government were initially those of excess demand and demobilization. The excess demand was the unavoidable consequences of the nature of the war effort. Full employment and long hours had generated high incomes, even allowing for high taxes, but these could not be spent owing to rationing and limited supplies. Hence at the end of the war there existed in both the household and the corporate sector a pent-up demand for restocking which threatened the 'boom and bust' that characterized the post-First World War period.[31]

At the same time, there existed around five million people in the armed forces, most of whom would quite quickly have to be reabsorbed into the civilian economy. Together, these two elements—the first predominant—provided a compelling case for the continuation of the wartime apparatus of controls into the period of peace. Politically, it would have been unthinkable for all this excess demand to be eliminated by tax increases. And in any case such a 'Keynesian' approach to the economy was far from dominant in Labour thought in 1945. The apparatus of physical planning chimed in much more closely with the predelictions of the dominant Fabian socialism, and more broadly with the rationalistic approach to the economy which embraced a much wider spectrum of opinion. Hence in the first two years under Labour, the economic surveys that were prepared in the Economic Section on the wartime national income basis were never published, and the one published in 1947 was much more in the vein of physical planning.[32]

Initially, fiscal policy under the Labour government was largely the unintended consequence of the run-down in public expenditure with demobilization, and the parallel reductions in tax rates. In fact, total tax revenue altered little, so most adjustment was on the expenditure side. The Budget swung quickly between 1945 and 1947 from deficit to surplus, despite a rough trebling of other forms of government expenditure under the same period.[33] Ministers were very conscious of the problems of inflation, but did not see the main weapon against this as fiscal policy via regulating demand. Not until 1947 did fiscal policy revert to something akin to the wartime 'Keynesian' budgets (see Chapter 6 below).

While planning fitted in with the ideological stance of labour, it was not the case that the government had a well developed programme for economic planning which it then proceeded to implement. Rather, planning, initially at least, combined an invocation of political will with, in large part, the structure of controls inherited from wartime, minus the controls over labour. But while the pressure of demand provided a rationale for the continuation of controls over consumption and imports for the time being, it was difficult to see such emphasis on determining (primarily) consumption patterns as a positive long-term programme. On the production side, labour controls were largely rejected on political grounds, even if manpower budgeting remained a perceived route to full employment.[34] Investment planning was pursued and indeed enhanced under the Labour government, most of this operating via controls of steel, timber, and other raw materials. But the main point of this was not tied to a clear policy of developing some sectors of industry at the expense of others, as notions of planning in the Soviet sense might suggest. Some of it was linked to a desire to favour some sectors—notably housebuilding—but most was related to the balance of payments problems that loomed so large across all economic policy-making.[35]

Overall, as the government's pre-eminent role in the economy was reduced, and as the pressure of excess demand and dollar shortage eased, planning came to be redefined. The emphasis moved away from the traditional socialist concerns with the physical regulation of output, which the Labour government never equipped itself to pursue, and towards national income planning plus regional policy.

Monetary policy did not play much of a role under Labour. The policy of cheap money, so successfully pursued during the war, was continued. The impact of this policy was much less than would have been the case because investment was regulated (held down) by direct controls, and therefore low interest rates did not determine investment levels. The budgetary reasons for low interest rates were compelling, given the scale of the national debt, though the attempt because of this to reduce the rate below 3 per cent was over-ambitious, and tended to rebound against the policy. However, the common complaint against the policy—that it led to inflation—has been rejected by the most recent comprehensive account of the Labour government's record.[36]

Though many of the individual economists involved left the civil service at the end of the war, the much greater role of economists and economic advice in the central bureaucracy continued into peacetime. Indeed, a new apparatus of economic work and advice was constructed in 1947 with the setting up of a Central Economic Planning Staff (CEPS). This was the closest Britain got in this period to having a central apparatus devoted entirely to planning, and it followed the publication in the same year of the closest thing to a full plan—the first Economic Survey of 1947. But this was not a plan in the sense of a programme of objectives, and mechanisms by which resources could be redeployed to achieve them. Rather, it was, like most planning in this period, concerned with sharing out resources whose supply was constrained by foreign exchange and other problems. It is symptomatic of policy development in the late 1940s that eventually the CEPS was absorbed into the Treasury. This was no longer 'a department like other departments, a Ministry of Finance when finance was no longer the key to economic policy'.[37] Rather, the areas where it had maintained its role—domestic and external finance—re-emerged as central after 1945, and with that the Treasury was restored to its place as the ministry of economics as well as of finance.[38] The Economic Section remained outside the Treasury (until 1953), but the trend was strongly in the direction of concentrating economic advice and policy-making in the Treasury, as the traditional financial instruments of policy increased in importance and physical controls were put on a 'bonfire'.

Of course, policy-making in this period was not dominated only by the issues of economic management sketched above: Labour's programme of extensive nationalization was also being carried out. The reasons for the nationalization are complex and range from an ideological hostility to private ownership to the much more specific concern to rebuild certain industries (notably, coal and railways), where the chance of such reconstruction under private ownership looked small. The impact of those nationalizations on the management of the economy was small—and the managers of the nationalized industries quickly emphasized their individual *raison d'être* against attempts to use them as instruments of economic policy (see Chapter 5 below).

Also important was the Development Areas policy. During the debates over the 1944 White Paper (see Chapter 3 below), structural-cum-regional policy was stressed as an *alternative* to demand management by the Treasury. The Board of Trade, by contrast, forwarded both instruments of policy, and the White Paper eventually embodied that view.[39] The Distribution of Industry Act of 1945 to a limited extent built upon the Special Areas Act of 1934, but its main impetus came from structures and policies built up during the war.

The wartime problem had been to mobilize as fully as possible the idle labour in the depressed areas, and to do this a successful policy of taking work to the workers was pursued. The policy involved close co-operation between the Board of Trade and private industry, and this was continued into the postwar period, when the conditions of excess demand (as well as building licensing) provided an ideal context for steering industry to the old industrial areas. The Act of 1945 focused on the provision of land, estates and buildings by public bodies to be used by private industry.[40]

In the conditions of buoyant demand, such policies proved successful in preventing a re-emergence of the regional problem on anything like the 1930s scale, though there remained areas, especially in south Wales and Lancashire, which in the early 1950s were commonly perceived as problem areas. As Booth emphasizes, the crucial breakthrough in this policy area was that, from the Act of 1945, 'the depressed areas would be treated as an economic and industrial rather than a social welfare problem'.[41]

Very significant for economic management was the construction of the welfare state. A central reason for the enhanced capacity of government in the twentieth century to manage the economy has been the growth of public expenditure (and the parallel growth of tax revenue). A large part of this growth has been on welfare expenditure. Thus, the welfare expansion of the late 1940s marked an advance in the capacity of government to manage the economy, which has proved a very enduring shift. While public expenditure fell back from its level of over 50 per cent of GNP during the war, it never sank to prewar levels: it was 30 per cent of GNP in 1950/1 compared with 16 per cent in 1938/9.[42]

The expansion of the welfare state raises again the question of how far this growth was inimical to Britain's economic expansion in the long run. How far is it true that the building of this 'New Jerusalem' was at the expense of rebuilding the British economy?[43] This raises a number of general issues which cannot be dealt with here—for example, how far in a direct sense welfare state expenditure might be functional to economic growth by its effects on health and education.[43] Equally of interest would be to know how certain northern European countries have been able to combine very comprehensive welfare states with highly successful economic management. Here the issue is looked at in a much narrower view: can the Labour government's economic policies be characterized as sacrificing the economic future to welfarism?

As noted above, the period saw a considerable success in reconstructing Britain's trading position. This emerged in part from a helpful international environment, but also from the clear priority accorded to exports within the government's policies. This is apparent in the picture of the overall pattern of resource used in the period. For the period 1946–52, the period of impact of Labour's policies, most of the growth of resources went into exports, absorbing nearly three-quarters of the growth of GDP. Most of the rest went into investment. Over the period, while GDP grew by 15 per cent, fixed capital formation rose by 58 per cent, and exports by 77 per cent, consumption rose by only 6 per cent.[45]

It is true that at various times under the Labour government measures were taken to restrict investment expenditure, but this was because of the desire to divert resources into exports, which in many areas were competitive with the claims of investment. And while it is the case that a substantial part of investment went into housing (as it always does), investment in plant and machinery dominated. If there was a clear diversion of resources away from investment (apart from to exports) it was to rearmament, after 1950.[46]

Cairncross's measured assessment of Labour's overall achievement fits very poorly with the picture painted by Barnett, and is worth quoting in full:

> Few governments also have held back consumption more assiduously so as to let the pace be set by exports and investment, as recommended by a later generation of experts on growth. They were successful in achieving a fast growth in exports, eliminating in turn the external deficit and then the dollar deficit and sustaining a high level of industrial investment in spite of the virtual cessation of personal savings. But they did not succeed in raising the rate of growth to the level that their European neighbours proved capable of maintaining. It must be very doubtful whether any set of government policies could have done more.[47]

The Conservatives, 1951–5

The pattern of economic policy in most areas was set for many years by the Labour government of 1945–51. Not only did the Conservatives generally maintain the welfare state policies and nationalizations that Labour had pursued, but they also took over the general trend of policy away from physical controls and towards a focus on budgetary policy, albeit with an ideological enthusiasm not always apparent under Labour.

Like the previous government, the government of 1951–5 faced periodic foreign exchange difficulties, but these had now much more the character of speculative bubbles on a reasonably healthy trend, rather than deep crises, as occurred in 1947 and 1949.[48] In consequence, domestic policy was less subject to the constraints that it had faced under Labour.

The immediate inheritance of the Conservatives was a balance of payments crisis, which in large part was an effect of the Korean war and the consequent shift in the terms of trade against Britain. Policy debate at the time tended to magnify this crisis as portending a collapse of the balance of payments position. This view was used to justify extremely radical proposals for going back to convertibility of the pound (abandoned after the brief episode in 1947), allowing the exchange rate to float, and blocking the sterling balances held in London. This proposal, code-named 'Robot', emanated from officials in the Bank of England and the Treasury, though it was opposed by the Economic Section. Though pushed by the chancellor of the Exchequer, the proposal was eventually dropped in favour of a sharpening of existing import-restrictive policies.[49]

This episode emphasized the vulnerability of the British economy to changes in the terms of trade. But this, of course, works both ways, and from late 1951 continually through the early 1950s, the terms of trade moved in Britain's favour, greatly easing the balance of payments position. This made possible the easing of import restrictions, so that by 1955 those still in place were largely residual in character, and not a substantial instrument of policy.[50]

The strengthening of the underlying balance of payments position did not prevent a powerful concern with the strength of sterling as a policy objective. This partly reflected the continuing proneness of the pound to difficulties because of low reserves, the scale of outstanding sterling balances, and the scale of outflow on capital account, both private and military. It also reflected a growing emphasis on the value of the currency not only as a basis for sterling's role as an international currency, but also as a measure of confidence in the government's policies. Some impact was felt on policy pronouncements by the related view that domestic policy's correctness or otherwise could be clearly measured by the balance of payments position. This implied a tight constraint on domestic policy because of the possible impact of such policy on the external account.[51]

Such a view was, along with the current balance of payments difficulties, a more fundamental reason for advocacy of the Robot plan. It would have meant (something akin to the policies of the mid-1980s) an attempt to use exchange rate levels as a central determinant of domestic policies, especially in the monetary field. The rejection of such an approach had broad significance, in that it signalled the Conservatives' commitment to a policy stance similar in character (if not always in rhetoric) to its predecessor.

This was perhaps most striking in the area of public expenditure and fiscal policy. Despite what was said at the general elections of 1950 and 1951, public expenditure continued to grow.[52] At the same time, fiscal policy was used, for the first time, to expand demand and reduce unemployment. Under Labour the problem had been to choke off excess demand, and the economy naturally 'disinflated' itself as output grew. But in 1953, faced with rising unemployment, Butler, the chancellor of the Exchequer, cut taxes to boost consumption demand. Indeed, given their strong desire to reduce taxes, the Conservatives showed a much greater willingness to cut than to increase taxes, even when macroeconomic conditions required the latter, and especially when an election loomed, as in 1955.[53]

The policies in those areas led to the coining of the term 'Butskellism' to suggest the similarity of the policies of Butler and those of the last Labour chancellor, Gaitskell. Certainly there were compelling reasons for similar policies in the initial period of Conservative government. The closeness of the 1951 election result, with Labour actually receiving more votes than the Conservatives, and the cross-party agreement on the need for rearmament, pushed policy on to similar tracks. And even in the longer run, the dismantling of controls—over investment by 1954, over most imports by 1955—fitted in with the trend of policy under Labour.[54]

Nevertheless, there were differences in economic ideology, and to some extent these were matched by policy changes. Most obvious is the revival of monetary policy. Though fiscal instruments remained dominant, there was from 1951 a use of Bank rate, of hire purchase controls, and of quantitative limits on bank lending to supplement fiscal policy. To some extent this was a logical corollary of the decline of physical controls, though it also fitted with traditional Conservative economic ideology as to the efficacy of such instruments and, in particular, to the importance of combating inflation by their use. One implication of this revival of monetary policy was the reassertion of the role of the Bank of England, which, despite nationalization in 1946, came to play once again a substantial role in policy-making.[55]

Butler himself, in comparing his policies with those of Gaitskell, argued that 'Both of us, it is true, spoke the language of Keynesianism. But we spoke it with different accents and with differing emphasis.'[56] One reason for this convergence was the realization that full employment, generated largely by the world-wide investment and trade boom, was here to stay. In this environment 'Keynesianism' ceased to be a revolutionary doctrine.

In 1951 the Conservatives had wholeheartedly committed themselves to the objective of full employment (see Chapter 8 below). Once the economy revived easily from the 1952 recession, this commitment looked secure. Not only the government, but the private sector also came to believe that there was to be no return to mass unemployment for the foreseeable future. For the private sector, this was probably important in the investment boom of the following years.[57] For governments, it meant a shift of emphasis in domestic policy away from employment both back towards the traditional concern with inflation, and towards the rather newer concern with growth.

As already noted, the containment of inflation had been an important policy preoccupation during the war, and more than any other single element was the reason for the Treasury's adoption of the new approach to budgetary policy. Parallel to this was the growth of concern with the inflationary effects on full employment.

Right from the beginnings of the discussion of the feasibility of a full-employment regime in the 1930s, it had been widely perceived that this posed potential problems for the behaviour of wages and prices.[58] One way of coping with this was seen to be by making any government commitment on employment conditional on restrained wage behaviour by trade unions—a kind of social contract. Such an approach was written into the 1944 White paper (para. 49), though this resulted as much from the failure to agree on any alternative policy as from any positive belief that such an approach would work.[59] There was general acceptance that the issue was an essentially political one, and there was widespread political resistance to any fundamental long-term departure from the tradition of free collective bargaining.

Under the Labour government, the first attempt was made at what was to become a hardy perennial of policy—an incomes policy. Under Cripps between 1948 and 1950 an agreement was made with the Trades Union Congress (TUC) which did see a fall in wage increases, though the policy was eventually undermined by the Korean war rise in import prices and the growth of perceived anomalies.[60]

As noted in Chapter 7 below, the Labour government continued its concern with wage inflation to the extent of proposing a White Paper on the subject in 1951, shortly before its fall from office. The Conservatives also saw the control of inflation as crucial, and this was one reason for the revival of monetary policy. But in fact, in the early 1950s there was a sharp slowdown in the rise in prices (see Table 1), to the extent that a respected commentator could speculate on whether the 'great inflation of 1939–51' was now at an end.[61] But by 1955 the steady upward pressure was re-asserting itself, and thenceforth was never to be far from the centre of policy concern.

Other things equal, governments had always seen growth in economy as desirable. But from the 1950s there was an increasing emphasis on growth as on explicit policy concern, especially because of the belief that such growth would act as a solvent of all kinds of political and economic issues. Butler, as chancellor of the Exchequer, held out the enticing prospect of a doubling of the standard of living every generation.[62]

Notes

1 R. S. Sayers, *British Financial Policy 1939–45* (London, 1956), 23–4; W. K. Hancock and M. Gowing, *British War Economy* (London, 1949), 46–62.

2 A. S. Milward, *War, Economy and Society 1939–45* (London, 1977), 40–1, 89.

3 Sayers, *British Financial Policy*, 5.

Table 1. Main Economic Indicators, 1939–55

	Unemployment (%)	Prices (1913 = 100)	Balance of payments Current balance (£m)		Terms of trade (1963 = 100; exports/imports)	GDP (1913 = 100, constant factor cost)
	(a)	(b)	(c)	(d)	(e)	(f)
1939	5.8	158	−250		141	128.9
1940	3.3	179	−800		125	141.8
1941	1.2	197	−820		123	154.7
1942	0.5	210	−660		139	158.5
1943	0.4	217	−680		132	162.0
1944	0.4	222	−660		139	155.6
1945	0.5	226	−870		134	148.8
1946	1.9	236	−230	−230	132	142.3
1947	1.4	249	−351	−381	123	140.7
1948	1.3	268	164	26	119	144.7
1949	1.2	275	153	−1	121	150.0
1950	1.3	283	447	307	112	154.9
1951	1.1	311	−326	−369	102	159.5
1952	1.6	338	163	163	111	159.2
1953	1.5	349	145	145	122	165.5
1954	1.2	355	117	117	120	172.2
1955	1.0	371	−155	−155	120	178.5

Sources: Cols. (a), (b), (c), (e), (f): C. H. Feinstein, *National Income, Expenditure and Output of the United Kingdom 1855–1965* (Cambridge, 1972); Col. (d): *Annual Abstract of Statistics*, Annual Supplement (1986).

[4] The emphasis on inflation also led to a policy of subsidizing the cost of living, as well as being one reason for physical planning, in order to prevent wages and prices from being bid up in the expanding war sectors.

[5] J. M. Keynes, 'How to Pay for the War', *Activities 1939–45: Internal War Finance* (Collected Writings, 22; London, 1978).

[6] Sayers, *British Financial Policy*, 69.

[7] D. N. Chester, 'The Central Machinery for Economic Planning', in D. N. Chester (ed.), *Lessons of the British War Economy* (London, 1951), 5–33.

[8] Hancock and Gowing, *British War Economy*, 452. On some of the complexities of planning see E. Devons, 'The Problem of Co-ordination in Aircraft Production' in Chester, *Lessons*, 102–21. On rationing see A. Booth, 'Economists and Points Rationing in the Second World War', *Journal of European Economic History*, 14 (1985), 297–317.

[9] D. N. Chester, 'The Central Machinery for Economic Policy', in Chester, *Lessons*, 6.

[10] A. Cairncross, 'An Early Think Tank: Origins of the Economic Section', *Three Banks Review*, 144 (1984), 50–9.

[11] D. MacDougall, 'The Prime Minister's Statistical Section', in Chester, *Lessons*, 58–68.

[12] Milward, *War, Economy and Society*, 130.

[13] J. Tomlinson, *British Macroeconomic Policy Since 1940* (London, 1985), Ch. 1.

[14] A. Cutler *et al.*, *Keynes, Beveridge and Beyond* (London, 1986).

[15] P. Addison, *The Road to 1945* (London, 1977).

[16] Ibid., Introduction and Ch. 10.

[17] C. Barnett, *The Audit of War* (London, 1986).

[18] At 1938 factor cost. C. H. Feinstein, *National Income, Expenditure and Output of the United Kingdom 1855–1965* (Cambridge, 1972), Table 5.

[19] Ibid. See also A. Cairncross, *Years of Recovery: British Economic Policy 1945–51* (London, 1985), Ch. 1, for a summary of the position of the economy at the war's end.

[20] Hancock and Gowing, *British War Economy*, 551.

[21] Feinstein, *National Income*, Table 37.

[22] Cairncross, *Years of Recovery*, 7.

[23] There is remarkable little of this kind of material in Barnett, given his argument.

[24] Hancock and Gowing, *War Economy*, 541.

[25] Cairncross, *Years of Recovery*, 63–4.

[26] T. Balogh, 'The International Aspect', in G. D. N. Worswick and P. H. Ady (eds.), *The British Economy 1945–50* (Oxford, 1952), 491–4.

[27] S. Strange, *Sterling and British Policy* (London, 1971).

[28] Cairncross, *Years of Recovery*, 74, 500, 503.

[29] R. N. Gardner, *Sterling–Dollar Diplomacy* (Oxford, 1956).

[30] Cairncross, *Years of Recovery*, 50; see also Ch. 9 below.

[31] D. H. Aldcroft, *The British Economy*. Vol. I, *The Years of Turmoil 1920–1951* (Brighton, 1986), 192–3.

[32] R. W. B. Clarke, *Anglo-American Collaboration in War and Peace 1942–9* (Oxford, 1982), xx–xxi, 77–9. Part 1 of the published Survey for 1947 was drafted personally by Cripps.

[33] Cairncross, *Years of Recovery*, 420–1.

[34] Ibid., 302.

[35] Ibid., 299–353.

[36] Ibid., 427–45. Also J. C. R. Dow, *The Management of the British Economy 1945–60* (Cambridge, 1965), 223–7.

[37] Cairncross, *Years of Recovery*, 50.

[38] There was a short-lived Ministry of Economic Affairs in 1947 under Cripps, but this was dissolved when he became Chancellor of the Exchequer in November 1947. It was revived, briefly, in 1950 under Gaitskell.

[39] The White Paper had a whole section on the 'Balanced Distribution of Industry and Labour' which explicitly played down the role of labour mobility as a primary method of solving regional unemployment.

[40] D. Jay, *Change and Fortune: A Political Record* (London, 1980), Chs. 5, 6, A. Booth. 'The Second World War and the Origins of Modern Regional Policy', *Economy and Society*, 11 (1982), 1–21.

[41] Booth, 'The Second World War', 16.

[42] Cairncross, *Years of Recovery*, 420.

[43] Barnett, *The Audit of War*, especially Chs. 1 and 12.

[44] Barnett is certainly not an advocate of *laissez-faire*, and his emphasis on the educational reasons for British economic decline would fit in with a massive expansion of (state) spending on education.

[45] Cairncross, *Years of Recovery*, 25–6.

[46] Ibid., 453–5. Compare Barnett, *The Audit of War*, 264.

[47] Cairncross, *Years of Recovery*, 500.

[48] This was, broadly, the Radcliffe view. See *Report of the Committee on the Working of the Monetary System*, cmnd. 827 (1959). Compare M. F. Scott, 'The Balance of Payments Crises', in Worswick and Ady, *The British Economy 1945–50*, 205–30.

[49] Cairncross, *Years of Recovery*, Ch. 9.

[50] Scott, 'The Balance of Payments Crises', 217; P. Streeten, 'Commercial Policy', in Worswick and Ady, *The British Economy 1945–50*, 86–90.

[51] Dow, *Management of the British Economy*, 67–70, 80.

[52] I. M. D. Little, 'Fiscal Policy', in Worswick and Ady, *The British Economy 1945–50*, Table 9.

[53] C. Kennedy, 'Monetary Policy', in Worswick and Ady, *The British*

Economy 1945–50, 302. For much more discussion of employment policy in this period, see Ch. 8 below.

[54] G. D. N. Worswick, 'The British Economy 1950–59', in Worswick and Ady, *The British Economy in the 1950s*, 15–16.

[55] Dow, *Management of the British Economy*, 69–70.

[56] R. A. Butler, *The Art of the Possible* (London, 1971), 160.

[57] Worswick, 'The British Economy 1950–59', 31.

[58] R. Jones, *Wages and Employment Policy, 1936–86* (London, 1987).

[59] Ibid., Ch. 6.

[60] Cairncross, *Years of Recovery*, 406–7.

[61] A. J. Brown, *The Great Inflation 1939–51* (Oxford, 1953).

[62] J. Tomlinson, *British Macroeconomic Policy since 1940* (London, 1985), Ch. 4.

3

The 1944 White Paper on Employment Policy: An Exercise in Damage Limitation?

Over the last twenty years, wide-ranging debates have occurred regarding the character and reasons for the policy responses to the mass unemployment in Britain in the 1920s and 1930s, or more often the seeming lack of responses.[1] What we can, by now, call the traditional approach to these matters emphasized the strength of resistence to, and hence slow permeation of, new economic ideas among policy-makers. Thus the slowness of change in policy was seen as mainly the consequences of the slowness in the penetration of Keynesian ideas.[2]

More recent work on the period has approached the matter in a different light. Rather than focusing on the extent of adoption of new economic ideas, evidence for which in the 1920s and 1930s looks very thin, this work has emphasized the political and administrative objections that central government (most notably the Treasury) had to some of the perceived implications of radical policy changes. While this approach does not ignore changes in economic theory, in understanding policy changes (or the lack of them) it puts greater emphasis on other aspects of the context in which economic policy-making takes place.[3]

Like most attempts at historiographical revision, this has met with a mixed reaction. One response has been to concede that there is indeed little evidence of changes in the theory adhered to by policy-makers in the 1920s and 1930s, but still to focus on this aspect as a crucial determinant of policy-making. These two points are reconciled by pushing forward the date of effective permeation of the new Keynesian theories into the early, or even late, 1940s.[4] This leads to an interpretation of the 1944 White Paper on Employment Policy as a sign of movement towards acceptance of Keynesianism, 'but falling far short of a complete Keynesian revolution . . . senior Treasury officials showed themselves wanting in their understanding of Keynesian economics at least as late as 1947.'[5]

That most Treasury officials were neither fully conversant with, nor adherents of, Keynesian theory in the period leading up to the 1944 White Paper is indisputable. What, however, seems more interesting is the question of whether the issue of intellectual allegiance is the best way of approaching the 1944 White Paper. In other words, cannot the reinterpretation of the interwar period be also applied to the early 1940s? In this light, the focus would be on the Treasury's perception of the political and administrative constraints under which it operated, and how far these were perceived as modified during the period up to 1944. This chapter, therefore, attempts to see how far the 'revisionist' approach can be justified by looking in detail at the making of the 1944 White Paper.

Early Wartime Discussions of Employment Policy

In the broadest sense, we can characterize the war period as generating what might be called a social democratic consensus on the need for widespread social reform, especially in the areas of education, housing, health care, social security, and employment.[6] Of these issues, the one that seemed to have most political salience in the middle of the war was that of social security, most notably following the publication of the Beveridge Report of 1942, and its author's successful promotion of that document.[7] The events surrounding the 1942 Report had two very important consequences for employment policy. In the first place, the issue of whether Beveridge's proposals could be afforded by postwar Britain pushed much more to the fore within the Treasury the issue of the likely shape of the economy in the postwar period. In particular, this led to work on the post-war national income, and on the plausibility of Beveridge's 'Assumption C', that full employment could be attained. Second, the knowledge in 1943 that Beveridge was himself preparing a report on full employment gave a political urgency to the formulation of government policy which culminated in the publication of the 1944 White Paper.[8]

Work on employment policy within Whitehall had gone on prior to Beveridge's report, notably in the Economic Section of the Cabinet office. Although in 1940/1 the Section's work was, unsurprisingly, dominated by urgent war-related tasks, at least as early as October 1940 Durbin was pressuring for work on reconstruction, and early in 1941 Jewkes was explicitly setting out an agenda for research on the feasibility of full employment.[9] After considerable discussion in the Economic Section, Meade produced the famous Memorandum, *The Prevention of General Unemployment*, in the spring of 1941.[10] This distinguished between frictional, structural, and general unemployment, but emphasized that the reduction of the latter was a precondition for the reduction of the other types. Thus the paper placed employment policy firmly in an income–expenditure framework. To combat what it saw as the likely prevalence of unemployment beyond the period of immediate postwar boom, the Memorandum recommended an expansionary monetary policy, a public works policy, possible stimulation of consumption by measures such as the relaxation of the terms of availability of hire purchase (HP) or tax reductions, and an expansionary budgetary policy.

On the question of budgetary policy, Meade immediately faced up to two of the issues that were later to become very important. He accepted that there would be a large wartime debt accumulated by the close of the war, and also that a budgetary policy tied to combating unemployment might itself lead to a continuous growth of debt. He accepted also that such an expansion of the debt could exert a depressing influence on business enterprise, and therefore he outlined the possibility of a capital levy, death duties, or a wealth tax as a way of financing debt interest out of taxes on capital rather than income.

Notably absent from the paper in the light of subsequent developments was any discussion of variations in social insurance contributions as a means of regulating consumption, or any discussion of the issue of 'confidence', except in the rather restricted sense noted above. Also of note is that a member of the Economic Section (Durbin) criticized the paper on the grounds that the emphasis on public works was 'an unnecessary and elaborate piece of ritual' if the Budget could be used to vary the level of consumption directly. Durbin saw the solution in such deficits 'to be covered by borrowing from the banks at a zero rate of interest'.[11]

Thus Meade's paper may be seen as an attempt to meet some well-known Treasury sensitivities. It took up the issue of budget deficits, albeit only in respect of their effects on the national debt. Meade did not discuss the possibility of money-financed deficits (although he had advocated this prewar, and was to do so again later), though they would seem perfectly compatible with the Keynesian notion of an unemployment equilibrium.[12] Equally, by emphasizing public works, the paper fitted in with the Treasury's qualified acceptance of such policies, at least on a small scale in the late 1930s.[13]

This initial Economic Section work entered policy discussions only with considerable delay, after the creation of a Committee on Postwar Economic Problems, itself seen as a complement to the discussions of the postwar international economy stimulated by the clauses of the Mutual Aid Agreement with the United States.[14] This committee considered, without any great sense of urgency, Meade's paper and a reply by Hubert Henderson.[15] This was dismissive of Meade's paper, not so much directly as by focusing on the immediate postwar transition period rather than the longer run, putting forward a decidedly pessmistic view of what that longer run might look like, and saying little specifically on possible policy options.[16] The Committee did not come down on either Meade's or Henderson's side, but wrote an eclectic report which was circulated to the Ministerial Reconstruction Priorities Committee.[17]

At the same time as these rather marginal discussions were occurring in the Internal Economic Problems Committee, the Treasury was considering budgetary policy. This was always of course a primary concern of the Treasury, in questions of employment, and the discussions of budgetary policy show how far the Treasury's concerns of the 1920s and 1930s, as taken up by Middleton, continued.[18] In May 1942 Henry Clay wrote to Hopkins arguing that too much emphasis was being put on public works despite their inefficiency. 'Of course it may be necessary, as a means of reassuring Conservative opinion, to concentrate expenditure on capital goods—just as in Lamb's fable it was necessary on religious grounds to burn down a house when you wanted roast pork. But the practice is in itself uneconomic, and an attempt might be made to educate opinion.' In reply, Hopkins stressed that, while he accepted much of Clay's views, including that fact that 'the difficulties about public works are chiefly practical', the crucial issue was that of budget deficits, whether arising from unremunerative public works or from a more direct method. To such deficits he was opposed, 'because it is useless to hope that the Chancellor would propose taxation to secure in good times budget surpluses corresponding to Budget deficits in bad times'.[19]

Hopkins was also opposed to the idea of a capital budget, as advocated by, for example, Keynes. His objection here was partly that direct capital investment by central government was very small, since most public sector investment was undertaken by local authorities, but more fundamentally that such a budget threatened deficits in disguise. 'I think we ought to avoid like the plague any Extraordinary Budget similar to continental ones in recent times. Such a device is merely an invitation to gloss over the facts of a difficult situation by putting into that budget, on one pretext or another, anything that should be, but cannot be, or is not currently paid for.'[20] These points bring out how far the Treasury's position remained one of emphasizing that the 'fiscal constitution' of balanced budgets acted as a constraint on the growth of expenditure by containing the natural demands of politicians and sectional interests for new expenditure. As the debate on employment policy moved closer to centre-stage from late 1942, this position remained a bedrock of the Treasury's views.

The First Rounds of Debate

This section concentrates on the main lines of objection by the Treasury to the proposals, largely emanating from the Economic Section, for policies to maintain employment after the war. Plainly these debates, as always with policy debates, were diffuse and complex. There was not, as Meade[21] has emphasized, a unitary Economic Section view opposed to a unitary Treasury view. However, the debates can be reduced to five major areas without doing too much violence to the historical record. (The issues of the inflationary impact of employment policy and its international ramifications are not dealt with here, as they are largely separate areas of debate which are dealt with elsewhere[22].) These areas are: (1) the nature of the likely unemployment problem; (2) control of investment (both public and private); (3) control of consumption via the level of social security; (4) control of consumption via tax devices; and (5) use of the Budget.

The agenda for discussion of postwar employment policy was set in 1943 by an Economic Section paper prepared at the behest of the Reconstruction Priorities Committee, entitled 'The Maintenance of Employment'.[23] The thrust of this was very similar to Meade's 1941 paper, focusing again on the post-transition period, and on the problem of general unemployment. One important difference to the 1941 paper was the advocacy of variations in social insurance contributions, on which considerable work has been done in the Economic Section since the discussion of Beveridge's 1942 Report.[24] After considerable manœuvring, the whole issue of postwar employment was referred to a Steering Committee, chaired by Hopkins.[25]

On the issue of the general character of the unemployment problem, the Treasury argued that the Economic Section's view 'is in important respects out of focus'. This criticism derived from the Section's alleged focus on general unemployment to the neglect of structural unemployment. In consequence, the Treasury wanted to emphasize the importance of industrial location policy as the way of dealing with this problem. However, the Treasury accepted that 'it is undoubtedly true that if general trading conditions are depressed, the problem of heavy concentrations of unemployment in particular districts is made thereby far more intractable . . . there is no question, however, as to the importance of maintaining aggregate demand at a high level, and of avoiding as far as possible general depressions of trade.'[26]

In responding to this Treasury document, Robbins argued that the Economic Section did *not* play down structural unemployment, and that the Section differed with the Treasury only if the latter was asserting that if all structural problems were solved all employment troubles would be at an end. He also stressed that there was a fundamental theoretical position being asserted in the Section's arguments, that the determination of saving by investment was crucial to the issue of unemployment.[27]

But, just as in 1931, before the Macmillan Committee, the Treasury did not rest its case on economic theory. Rather, Eady, on behalf of the Treasury, stressed the political and administrative problems of converting the broad principles of aggregate demand theory into policy. In this, the Treasury were

> bound to make certain assumptions as to the world in which they were framing their policy, assumptions as to the continuance of a large part of

the field of industry as free enterprise, as to the relations between the central government and local authorities, as to the legislative and administrative problems involved, and also as to the problem of securing some continuity in policy for a period under Governments which might differ in political complexion.[28]

As with the Keynes/Hopkins clash before the Macmillan Committee, while undoubtedly different concepts of the functioning of the economy were at work, in important respects the disputants were talking past each other. The economists typically saw the issue in doctrinal terms, whereas the Treasury saw it as part of a complex policy environment in which doctrinal issues should be subordinate to calculations of political and administrative possibility. Hence the most usual epithet hurled at the Economic Section was that its arguments were 'abstract' and 'academic', which was only in part a code for 'theoretically unacceptable'.

The second major area of dispute was over the control of public and private investment. In the case of public investment, the Treasury focused not on the principle of public works, but on the characteristics of the postwar period. They argued that in that period the pent-up demand for public investment would be such that it would be infeasible to restrict it in the manner in which, they alleged, the Economic Section was suggesting. More specifically, they argued that the capacity of the Treasury to exert positive, that is, expansionary, influence on public investment, most of which was carried out by local authorities and public utilities, was severely constrained. They argued that financial inducement to such bodies to expand investment in the slump would face difficulties, because for example the bodies might postpone investment until such inducement was offered. As a result, the Treasury thought that to stabilize public investment was the best that could be hoped for, rather than to use it counter-cyclically.[29]

In reply, the Economic Section reasserted the importance of using public investment contra-cyclically. They attacked the negativism of the Treasury, but accepted that the absence of direct positive controls by central government over much public investment was a great weakness. However, they suggested that it should be possible to think of devices to influence at least some of this investment expenditure.[30]

The discussion of this issue focused on such administrative

practicalities, and it was agreed that a scheme whereby grants for new enterprises varied by some 'objective criteria' would be feasible. Strikingly, no theoretical issue of 'crowding-out', of the notorious Treasury View of 1929, entered the discussion either here or, it seems, anywhere else in the discussions of this period.[31]

The discussion of private investment has the curious feature that what has sometimes been regarded as a Keynesian policy—cheap money—was seen as much more problematic, as a policy for the postwar period, by the Economic Section than by the Treasury. The reason for this was that the good market economists of the Economic Section were unhappy with too long a period of controls, which they saw as the only alternative to the use of interest rates to damp down the level of investment demand.[32]

The Treasury also disputed the Economic Section's case for a Public Finance Corporation, disputing the existence of any financial gap, except perhaps for small firms. Here the difference between the protagonists was small, with the Economic Section also pressing the case for some public role in financial provision largely because they disliked the monopolizing impulses of private financial institutions in the face of declines in demand.[33]

More important was the issue of tax incentives for stimulating private investment. Here the Treasury called in aid the Inland Revenue, which argued that such tax privileges would conflict with the 'fair sharing out of the bill of national expenditure'. More specifically, the Inland Revenue pointed out that such tax privileges could accrue only to those who made profits, and also that it would of necessity be expensive, because the tax system could not discriminate between investment that would and would not have taken place in any case. While not wholly accepting the Inland Revenue's view of the basic principles of taxation, the Economic Section conceded the force of the other points, and instead advocated at least the examination of a subsidy system, which would deal anyway with the Inland Revenue's point about the limiting of tax privileges to profitable enterprises.[34]

The most detailed policy proposal in the Economic Section's paper was for a system of variations in national insurance contributions according to the level of unemployment. Against this scheme the Treasury argued two main points. First, such an automatic scheme would not discriminate between unemployment caused by structural dislocations and that caused by a general deficiency of aggregate demand. The Economic Section agreed that this problem would prevent the scheme from being introduced in the immediate postwar transition period, but the Treasury argued that the problem might be more general and more prolonged. Second, the Treasury argued that the calculation of an average rate of unemployment around which the variations in contribution would take place would be difficult, and hence such variations would carry a high risk of rendering the national insurance fund insolvent.[35]

This negative response triggered the most lengthy of the Economic Section's counterblasts to the Treasury. The Section stressed the need for a general and quick-acting stimulus to demand in time of depression—and in particular the problems of using public works at such times. 'For it is by no means certain that a sufficient volume of public works expenditure can be readily varied in timing, and it is virtually certain that such expenditure cannot be very rapidly and promptly turned on and off in response to the needs of the general economic position.'

In replying to the Treasury's objections to social insurance variations, the Economic Section pursued a rather curious tactic. Not only did they concede the problem of such measures not discriminating between structural and demand deficient unemployment, but they added their own difficulty: that if wage inflation were occurring at below full employment, this would be a reason for not expanding aggregate demand. Thus, 'other indices besides that of the unemployment percentage are relevant for decisions about the need for expansive or restrictive economic and financial policies'. But in the case of social security contributions, they argued, these considerations must be disregarded because of the need to balance the insurance fund.[36]

This rather curious argument seems to have been based on the Economic Section's trying to counter the well-known Treasury view against discretionary budget deficits (see below) by advocating an automatic mechanism. But such advocacy ran up against another Treasury view, that chancellors must have full discretion in their financial policies. This was not inherently contradictory, once it is noted that the Treasury's view was for discretion *within* a position of budget surplus. This was the rock on which the Economic Section's scheme eventually foundered.[37]

Alongside advocacy of variation in social security contributions to regulate the level of consumption, the Economic Section also argued for the continuation of the wartime tax credit (i.e. compulsory saving) system. This was coupled with an emphasis on the difficulty of using variations in tax directly to influence demand, because of the annual basis of tax changes, their slow effects, and also the possibility that postwar tax thresholds might rise sufficiently to exempt large numbers of wage-earners.[38]

The Treasury case was not to advocate such a use of the tax system. But in criticizing the Economic Section's discussion of tax privileges for private investment, the Inland Revenue also criticized the scheme of tax credits in peacetime. They argued that such a scheme would 'place a millstone round the Chancellor's neck in determining taxation policy from one year to the next'. The Inland Revenue was generally hostile to the use of taxation to regulate demand, but they argued that if this were to be done, then it could be done much more simply by direct changes in taxation than by a complex system of tax credits.[39]

This, as noted above, was a point made by Durbin back in 1941, and it is a curious feature of the debate that the Economic Section invested so much effort in variations in social security contributions, while making a relatively feeble case against direct use of taxation. Although it is not explicit, the reason for this seems to have been a calculation by the Section that deficits in the social security fund would not arouse quite the hostility of deficits in the ordinary Budget. In the same way, a tax credit system could be seen as constraining the discretion of the government over the ordinary budget; that is, high expenditure in a slump, via release of such credits, could be presented as financed by their accumulation in previous periods. Such political considerations could fit in with the further suggestion that the insurance fund itself could be split into two, with an ordinary, always balancing, section and a fund that could borrow in bad times to be repaid in good times from the ordinary fund.[40]

The final area of Treasury hostility to the Economic Section's proposals was that of budget deficits. Over half of the Treasury's response in 1943 was taken up with budgetary issues. The clear hostility to proposals for balancing the Budget over the trade cycle rather than annually was argued on a number of grounds.

First, there was the argument that in the immediate postwar period there would be both excess demand and a call for reductions in wartime tax levels, and this was seen as making difficult the accumulation of surpluses against future declines in demand. More generally, the Treasury saw the need for budgetary balance in the face of a stagnant or declining population with its implications for the tax base and hence the financing of the national debt. More broadly still, the Treasury emphasized the importance of confidence. Unlike during the war, in peacetime 'it is necessary to assume that a large part of our economic activity will remain dependent for a long time to come on the decisions and initiative of private individuals, and furthermore will be subject in some degree to the influence of external opinion'.[41]

The Treasury asserted that they did not rule out the Budget taking into account employment conditions, and indeed that it had not been ruled out in traditional budgetary practice. The implications of this was that taxes might not have to be raised in a slump, but that deliberate tax cuts or expenditure increases in such circumstances would be a different matter. This would create the necessity for later compensatory tax increases, which would be politically unpopular. 'The Treasury doubt whether it would be possible in practice to reconcile the application of this principle with the governing object of balancing the budget, with a margin to spare over a period of years.'[42]

Finally, the Treasury turned to the issue of a capital budget (though this was not in fact advocated by the Economic Section). A major point made in this context was that such a splitting of the Budget would facilitate the extension of the definition of capital to cover more and more expenditure, paving the way to 'extraordinary budgets' as widely practised on the Continent between the wars. The one reform they accepted was the publication of more data on the overall investment position, with perhaps a capital survey as an accompaniment to the established budgetary publications.[43]

In responding to these criticisms, the Economic Section accepted that the dangers existed. On the problems of the taxation, consequences of continued deficits, and a rise in the national debt, they argued that this might lead to a more serious examination of the possibilities of financing the public debt by the issue of non-interest-bearing loans and deposits rather than by the issue of interest-bearing securities.

On the disciplining of public expenditure by the balanced budget rule, the Section relied, rather ingenuously, on asserting the common sense of the British people, and their ability to realize that temporary deficits did not mean any uneconomic use of public money. Relatedly, on the issue of confidence, the Section believed that the analogies with 1931 were misplaced, that the populace was now better educated and could be more so. In addition, they advocated the retention of exchange control to guard against manifestation of loss of confidence via capital flight.[44]

In discussion of these issues by the Steering Committee, Eady accepted that the Economic Section was more cautious in its advocacy of financial heterodoxy than the Treasury response suggested. The target of the response, he said, was less cautious views, which he detected in various quarters, including the Board of Trade's own response to the Treasury's views.[45]

The Board of Trade's paper is important for the light it throws on the complexities of the debate. Its very existence in the arguments serves to emphasize, that, while picturing these as 'the Treasury versus the Economic Section' is not wholly misleading, neither is it the whole truth. More specifically, the contents of the paper throw a useful light on the way in which allegedly opposed arguments could be combined. The Board of Trade, reflecting a major aspect of its rather diverse set of departmental concerns, strongly asserted its support for the Treasury's emphasis on structural and local unemployment.[46] But the paper went on to say that this did not involve agreement with the Treasury's down-grading of the significance of aggregate demand.

> The higher this aggregate demand, the greater will be the mobility of labour out of depressed areas and at the same time the more inclined will industrialists be to move to these areas where unemployed labour is still available. In this most important sense, a policy of dealing with structural unemployment is, therefore, not competitive with, but complementary to, a policy of maintaining aggregate demand.[47]

The paper then proceeded to launch a wholehearted defence of budget deficits as instruments of macroeconomic management, in particular arguing that the threat of inflation from such deficits was small as long as unemployment persisted, and that confidence could be maintained by an education of the public in the principles of such a use of the budget. They too floated the possibility of financing the deficits through borrowing from the banks, and hence avoiding the problem of an accumulating national debt.[48]

This stance is important in making more difficult the assumption that the debate between the Economic Section's emphasis on aggregate demand, and the Treasury's on structural problems, was based on an underlying doctrinal disagreement. The Board of Trade could happily combine both positions—not because it was schizophrenic, but because this involved no logical contradiction, and seemed to provide a coherent answer to one of the Board of Trade's most pressing peacetime concerns: regional unemployment.

Towards the White Paper

The Steering Committee, as already noted, was chaired by Sir Richard Hopkins. He had taken a very hostile view of the Economic Section's paper, typically denouncing it as 'academic' among all its other faults. Nevertheless, the Report of the Committee leant towards the Economic Section's emphasis on aggregate demand as the starting-point for analysis. This seems to have been based on Hopkins's view that what ministers were particularly concerned with at this time was the country's capacity to afford the Beveridge proposals, and for this what mattered was the long-term employment prospects and not the problem of the transition period, which so worried the Treasury.[49]

Even so, the Report contained considerable qualification to the emphasis on aggregate demand, particularly by noting the problems of structural unemployment, arising especially in the export trades. 'It is clear, what has been said, that aggregate demand is not simply a tap which can be turned on or off, according to the state of trade, and which will provide an infallible remedy for all forms of unemployment.'[50]

Rather different views existed as to what had been at stake in discussions leading up to the Steering Committee's Report. Eady from the Treasury said that 'The whole discussion seemed to us very unreal, the main interest of Gaitskell and the Economic Section being to get an admission from the Treasury that the interests of employment would not be sacrificed to a rigid adherence to balanced budgets. We cheerfully gave such an assurance with the necessary qualifications . . .'[51] (The qualification referred to the issue of the national debt and confidence, as outlined above.)

One of the secretaries of the Committee, D. N. Chester,

expressed the view that a lot of the disagreements in the Committee's discussions 'arise not so much from difference of opinion as to the need for a particular measure on theoretical grounds, but from the view taken by the individual of the administrative or political practicality of the scheme'.[52]

If one takes Chester's summary as representing how the Treasury saw the issues, then both these viewpoints seem accurate. In other words, the two major protagonists were largely not fighting on the same terrain. The differences were obscured to some extent in the Report by the emphasis on the long term. Hence, when Keynes commented on the Report that it was 'an outstanding State Paper which, if one casts one's mind back ten years or so, represents a revolution in official opinion',[53] he was being rather optimistic, as well as putting a typically exaggerated emphasis on the significance of the state of official opinion.

This is illustrated by the discussions that ensued after the Steering Committee's Report and leading up to the 1944 White paper. The most intransigent opposition to the Economic Section's approach came from Hubert Henderson, who saw the Report under sway of 'crude sweeping abstract assertions which miss the crucial points', and engaged in an acrimonious dispute with Keynes in early 1944.[54] More significant from the point of view of the eventual form of the White Paper was Hopkins's view that for such a public statement there would need to be more qualification to the aggregate demand approach, and more on the transitional period.[55]

These qualifications once again ranged over all the fields of the previous debate. There was a continuing opposition to the specific proposals for the use of public works and variations in social insurance contributions as central instruments of demand regulation. Gilbert emphasized that the former were 'unfamiliar and uncertain'.[56] Eady protested against a draft of the White Paper that it 'should contain two major proposals, on the variation of public expenditure and on the variation of insurance contributions, which all the officials concerned believed impracticable or at best far less effective for their designed purpose than is claimed for them'.[57]

Hopkins recorded his attraction to the social insurance variation scheme, but then went on to say both that its applicability in some circumstances was doubtful, and 'that one cannot in practice expect successive Governments in the long term necessarily to act always in the face of public pressure, on grounds of wisdom rather than of political expedience'.[58] On this issue, as on others, the final White Paper expressed a hurried compromise which concealed the extent of the continuing differences. The scheme was advanced in the White Paper, but with a strong qualification as to the timing of its introduction, which in effect allowed the Treasury to prevent its ever being brought into existence.[59]

The Treasury, as always, saw itself as pushed into undesirable paths by political pressures. Thus, one strong ministerial pressure was for inclusion in the White Paper of some commitment to variations in tax as an instrument for varying the level of consumption. In responding to these pressures, Hopkins wrote:

> All these devices for operating through taxation were looked at by the Steering Committee with the help of Sir Cornelius Gregg and all alike were condemned. In a perfect world we ought to continue to condemn them now, but the real problem is how to satisfy a body of Ministers who have not time to study the whole implications of those very difficult questions, and also to satisfy a public opinion which is even less informed, by references which will have no ill effect as regards budgetary stability and are unlikely to be embarrassing to future Governments, who will probably not actually adopt any device of this kind.[60]

This quotation shows not only the typical Treasury concern for 'damage limitation' in the face of political pressure, but also its continuing focus on the budget deficit as the central issue of the full-employment debate. Henderson and Hopkins continued to express their belief that the proposals of the White Paper should stress the role of confidence in relation to any essaying of the possibility of budget deficits.[61]

The Significance of the White Paper

The significance of the 1944 Employment White Paper lay in the political commitment it embodied. Both ministers and officials perceived that politically some kind of commitment on employment was unavoidable, given the overwhelming desire for no postwar return to the 1930s, crystallized by the war's effects in virtually eliminating unemployment.

How far the compromises of the White Paper should be seen as a staging post on the road to a conversion of the authorities, notably the Treasury, to Keynesian theory is a moot point. As suggested above, it is possible to see the disputes as at least as much about the shifting political, administrative, and economic constraints within which policy should be made as with the theoretical specification of the economy's functioning. Meade seems to have recognized this, for example, in his attempts to make his social insurance scheme palatable to Treasury prejudices—albeit, in this case, to no avail. Thus he got to grips with the way the Treasury viewed the world in a manner that Keynes often did not—in, for example, his evidence to the Macmillan Committee.[62] This of course did not mean that the Economic Section and the Treasury were of one mind. The White Paper of 1944 *was* a compromise. But the burden of this chapter is that the compromise was more between different calculations of the administrative, political, and economic constraints likely to exist in the postwar world then between different types of economic theory.

Finally, it is worth emphasizing that in the longer run *none* of the remedies fought over in 1942–4 was actually deployed postwar. On public works, the Treasury line seems to have received some vindication from the difficulties encountered in varying postwar public investment.[63] The possibility of variation of social insurance contributions was written into the 1946 National Insurance Act, but was never brought into force. Direct control over private investment was limited to rather marginal tax devices. Eventually, it was variation of general taxation that came into use as the main instrument of economic management, something strongly resisted by the Treasury before 1944. However, for twenty years these tax variations were in the context of budget *surpluses* and very small overall public sector deficits,[64] so that crucial issues in the debates of the early 1940s, such as the effects of deficits on confidence and the national debt, never became vital policy issues until the 1970s.

Notes

1 See Chapter 1 above.

2 M. Stewart, *Keynes and After* (Harmondsworth, 1967); D. Winch, *Economics and Policy* (London, 1972); S. Howson and D. Winch, *The Economic Advisory Council 1930–39: A Study in Economic Advice during the Depression* (Cambridge, 1977).

3 J. Tomlinson, 'Why Was There Never a Keynesian Revolution in

Economic Policy'?, *Economy and Society*, 10 (1981), 72–87; R. Middleton, 'The Treasury in the 1930s: Political and Administrative Constraints on the Acceptance of the New Economics', *Oxford Economic Papers*, 34 (1982), 48–77; R. Middleton, *Towards the Managed Economy: Keynes, the Treasury, and the Fiscal Policy Debate of the 1930s* (London, 1985).

[4] G. C. Peden, 'Sir Richard Hopkins and the "Keynesian Revolution" in Employment Policy 1929–45', *Economic History Review*, 36 (1983), 281–96; A. Booth, 'The "Keynesian Revolution" in Economic Policy Making', *Economic History Review*, 36 (1983), 103–23; A. Booth, 'The "Keynesian Revolution" and Economic Policy Making: A Reply', *Economic History Review*, 38 (1985), 101–6.

[5] Peden, 'Sir Richard Hopkins', 296. Also Booth, 'The "Keynesian Revolution" ', 116, 123. See also R. J. Macleod, 'The Development of Full Employment Policy 1938–45', D. Phil. dissertation (Oxford, 1978). On Keynes's personal relation to the White Paper proposals, see T. Wilson, 'Policy in War and Peace: The Recommendations of J. M. Keynes', in A. P. Thirlwall (ed.), *Keynes as a Policy Adviser* (London, 1982), 48–63.

[6] P. Addison, *The Road to 1945* (London, 1977).

[7] J. Harris, *William Beveridge: A Biography* (Oxford, 1977), Ch. 17.

[8] Ibid.

[9] A. K. Cairncross, 'An Early Think Tank: Origins of the Economic Section', *Three Banks Review*, 144 (1984), 50–9, for the early Economic Section; PRO T230/12 and T230/13, *Discussion Papers of the Economic Section* (1940, 1941).

[10] PRO T230/13 EC(5)(41)22 (Second Revise), *The Prevention of General Unemployment*, Memorandum by Mr Meade.

[11] PRO T230/13, *The Prevention of General Unemployment*, Note by Durbin (27 March 1941).

[12] J. Meade, *An Introduction to Economic Analysis and Policy* (London, 1937); PRO T230/67, J. Meade, *A Plan for the Control of National Expenditure and National Income* (3 December 1943); also J. Buchanan and R. Wagner, *Democracy in Deficit: The Political Legacy of Lord Keynes* (London, 1977), 31–4.

[13] Howson and Winch, *The Economic Advisory Council*.

[14] PRO CAB 87/54 IEP(4)9, Minutes of the *Committee on Postwar Internal Economic Policy*.

[15] PRO CAB 87/55 IEP(42)21, Memorandum of the *Committee on Postwar Internal Economic Problems*, 'Post War Relation between Purchasing Power and Consumer Goods' (26 May 1942).

[16] Ibid., IEP(42)22, 'Assumptions of Long-Run Planning' by Economic Section.

[17] PRO CAB 87/56 IEP(42)26, Revise of 'Assumptions of Long-Run Planning'.

[18] Middleton, 'The Treasury in the 1930s'; PRO T160/1407 F18876, H.F. Division of the Treasury, *Postwar Budgetary Policy. Form of the Budget. Capital Budget Papers* (1941–4).

[19] Ibid. (1 May 1942, 13 May 1942, 22 May 1942). Meade responded positively to a copy of Clay's letter of 1 May 1942, but argued that social security variation was superior to general tax measures: T230/69 Meade to Clay (4 May 1942, 18 May 1942).

[20] Ibid., Hopkins, *Postwar Budgetary Policy* (16 September 1942).

[21] Correspondence with author, 26 July 1985. See also the interesting arguments of A. Booth, 'Simple Keynesianism and Whitehall', *Economy and Society*, 15 (1986), 1–22.

[22] R. N. Gardner, *Sterling–Dollar Diplomacy* (Oxford, 1956); R. Jones, *Wages and Employment Policy 1936–86*, (London, 1987).

[23] PRO CAB 87–13, PR(43)26 (18 May 1943).

[24] PRO T230/14, *Variations in Social Insurance Contributions*: J. M. Keynes, *Activities, 1940–46: Shaping the Post War World: Employment and Commodities* (Collected writings, 27; London, 1980, 207–19).

[25] PRO T161/SS2098, *General Papers Leading up to Report of Steering Committee on Postwar Employment.*

[26] PRO CAB 87/63 EC(43)6, *The Maintenance of Employment: Prefatory Note by the Treasury* (no date), 1, 2.

[27] PRO CAB 87/63 EC(43)11. *Minutes of Committee on Postwar Employment* (1 November 1943).

[28] Ibid.

[29] PRO CAB 87/63, Prefatory Note, 4–6; Treasury, *Planning and Timing of Public Investment*, Note by Economic Section (no date).

[30] PRO CAB 87/63 EC(43)14, *Planning and Timing of Public Investment*, Note by Economic Section (20 October 1943).

[31] PRO CAB 87/63 EC (43)12 (3 November 1943).

[32] Booth, 'The "Keynesian Revolution" and Economic Policy Making', 105; PRO CAB 87/63 EC(43)10, *Control and Timing of Private Investment*, Note by the Economic Section (19 October 1943).

[33] Economic Section, Control and Timing, ibid.; CAB 87/63. *Control and Timing of Private Investment*, Treasury (no date).

[34] Treasury, *Control and Timing*; also *Note by Inland Revenue* appended. See also CAB 87/63 EC(43)13, 14 (3 and 4 November 1943). The Inland Revenue always took a very conservative view of the role of tax policy. R. S. Sayers, *British Financial Policy 1939–45* (London, 1956), e.g. 101.

[35] PRO CAB 87/63, Treasury, *Control and Timing of Public Investment* and *Variation of Social Security Contributions.*

[36] PRO T230/14 EC(S)(42)18, *Variations in the Rate of Social Security Contributions* (21 July 1942), paras. 11–15.

[37] PRO T230/105, *National Insurance: Variation of Contributions.*

[38] PRO T230/14 EC(S)(42)18, *Variations*, paras. 4–6.

[39] *Note by Inland Revenue*,2.

[40] PRO T230/105, *National Insurance: Variation of Contributions*. Memo by Dow (2 May 1946). Meade had worked an automatic stabilizers before the war. J. Meade, *Consumers Credits and Unemployment* (Oxford, 1938).

[41] CAB 87/63, Treasury, *Prefatory Note*, 8.

[42] Ibid., 10.

[43] Ibid., 12–13.

[44] PRO CAB 87/63 EC(43)9, *Maintenance of Employment*, Note by the Economic Section (18 October 1943).

[45] PRO CAB 87/63 EC(43)11 (1 November 1943); also T160/1168/SS2098, Eady to Barlow (12 October 1943).

[46] A. Booth, 'The Second World War and the Origins of Modern Regional Policy', *Economy and Society*, 11(1) (1982), 1–21; D. Jay, *Change and Fortune: A Political Record* (London, 1980), Ch. 6.

[47] PRO CAB 87/63 EC(43)12, *Maintenance of Employment: Board of Trade Observations on Prefatory Note by the Treasury* (20 October 1943), para. 3.

[48] Ibid., para. 9(a). R. F. Kahn and W. B. Reddaway were both active Keynesians working in the Board of Trade.

[49] Peden, 'Sir Richard Hopkins', 289–90; PRO CAB 87/63, *Report of Steering Committee on Postwar Employment* (10 January 1944), para. 8.

[50] Ibid., para. 18.

[51] PRO T161/1168/S52098, Eady to Barlow (12 October 1943).

[52] Ibid., Chester to Hopkins (24 October 1943).

[53] Keynes, *Shaping the Postwar World*, 364.

[54] PRO T161/S52099.

[55] Ibid., Hopkins to Chancellor of Exchequer (28 March 1944).

[56] Ibid., Gilbert to Hopkins (21 April 1944).

[57] Ibid., Eady to Brook (26 April 1944).

[58] T161/1168/S42099/01, Hopkins to Chancellor of Exchequer (29 April 1944).

[59] PRO T230/105, *Variations in Social Insurance Contributions*.

[60] T161/1168/S52099/01, Hopkins (10 May 1944).

[61] T161/1168/S42099, Henderson to Eady, *Employment White Paper* (20 April 1944); Hopkins to Chancellor of Exchequer (2 March 1944).

[62] Reprinted in J. M. Keynes, *Activities 1929–31: Rethinking Employment and Unemployment Policies* (Collected Writings, 20; London, 1981), 38–157.

[63] J. C. R. Dow, *The Management of the British Economy 1945–60* (Cambridge, 1965), Ch. 8.

[64] R. C. O. Matthews, 'Why Has Britain Had Full Employment Since the War?', *Economic Journal* 78 (1968), 555–69.

4

The White Paper of 1944 and Beveridge's *Full Employment in a Free Society:* One Policy or Two?

Chapter 3 has attempted to show how the 1944 White Paper emerged from a compromise between the political pressure on the government to produce a commitment to full employment, and the Treasury desire to defend its traditional concerns. This chapter attempts to widen the discussion of that White Paper by comparing it with Beveridge's *Full Employment in a Free Society*.[1] This is especially appropriate as the records show not only that the imminent publication of Beveridge spurred the government to produce the White Paper, but also that the producers of the White Paper saw Beveridge's effort as the main competitor in proposals for employment policy.[2]

While in a very broad sense these two documents could both be said to mark a new phase in economic thinking and policy, the relation between them is unclear. Winch[3] has called Beveridge's book 'a far more radical document . . . a report which in terms of diagnosis and remedies was closer to the spirit of the Keynesian revolution than the White Paper'. By contrast, Jewkes, a contributor and strong supporter of the White Paper, thought Beveridge's proposals 'dangerous for freedom'.[4]

Certain broad themes are plainly similar in both documents. In his contemporary analysis of them, Meade argued that 'the most striking feature is the great similarity in the general treatment of the problem in the two documents'.[5] Above all, he emphasized their focus on maintaining aggregate demand as the starting-point for effective employment policy, in both cases coupled with policies on the location of industry and the mobility of labour. Plainly, it is in their emphasis on maintaining aggregate demand that these documents may be seen both as marking a new phase in policy discussion, and as being fundamentally similar. While such similarity is hardly to be dismissed, its implications for policy are extremely unclear; and it is the *ways* in which it was envisaged that demand might be expanded and stabilized that will be focused on here.[6]

The Budget

As has often been noted, the White Paper was profoundly ambiguous on the question of the possibility of incurring budget deficits for the purposes of demand management. On the one hand, 'None of the main proposals contained in this Paper involves deliberate planning for a deficit in the National Budget in years of sub-normal trade activity' (para. 74); on the other, 'in controlling the situation . . . the Government will have equally in mind the need to maintain the national income, and the need for a policy of budgetary equilibrium' (para. 79).

As we have seen in Chapter 3, the reasons advanced for not regarding the Budget are simply a tool of economic management were essentially twofold. The worry of the White Paper was the long-run growth of the public debt, which would be harmful by its effects on the level of taxation and its effects on confidence. On taxation, the White Paper notes the view that interest payments on the national debt are simply a transfer payment, imposing no overall burden on the community. But, it is urged, 'the matter does not present itself in that light to the tax payer, on whose individual effort high taxation acts as a drag' (para. 78). In addition, the White Paper argues, 'Both at home and abroad the handling of our monetary problems is regarded as a test of the general firmness of the policy of Government. An undue growth of national indebtedness will have a quick result on confidence . . . confidence in the future which is necessary for a healthy and enterprising industry' (para. 79).

By contrast, Beveridge's proposal is for the unambiguous subordination of budgetary policy to economic management. 'But once the possibility of deficient private demand is admitted, the State, if it aims at full employment, must be prepared at need to spend more than it takes away from the citizens by taxation, in order to use the labour and other productive resources which would otherwise be wasted in unemployment' (para. 182; also paras. 197–8).

And yet, matters are not quite so straightforward: 'Nevertheless, there are good reasons for meeting State outlay, so far as is practicable, from current revenue raised by taxation, rather than by borrowing' (para. 199). These reasons are twofold, and are quite strikingly different to the grounds expressed in the White Paper. 'The main reason is the objection to increasing the numbers and wealth of rentiers, that is to say of people with legal claims against the community entitling them to live at the cost of the community of the day without working, though they are of an age and capacity to work' (para. 199). For this reason, Beveridge supports a policy of cheap money, and cites with approval Keynes's famous call for the long-run 'euthanasia of the rentier'.[7]

Beveridge's other grounds of opposition to budget deficits and the growth of public indebtedness were from a rather different ideological slant. 'A subsidiary reason is that borrowing to meet State expenditure, in place of meeting it from current revenue, enables the Government of the day to avoid the unpopular task of taxing and the loss of votes from this unpopularity. In other words it increases the opportunities of general political bribery' (para. 199). This argument, though suppressed in the White Paper, was one that the Treasury had long seen as important, and was deployed in the arguments leading up to the White Paper.[8] It is also of course a major plank in recent attacks on 'Keynesianism'.[9]

Interesting as these two points are, they are less important than the absence of opposition to budget deficits on the grounds advanced by the White Paper—the effects of taxation, directly and (via confidence) indirectly, on investment. In fact, Beveridge does mention the first point, very much in passing, when he says there are grounds for keeping taxation as high as possible 'without stifling desirable enterprise' (para. 199). He argues that in fact tax rates will *not* have to rise to finance likely rises in national debt, basing himself on Kaldor's argument in Appendix C of the book (para. 198). More importantly, he explicitly attacks the confidence argument. In commenting on the White Paper at the end of his own book, Beveridge wrote: 'The Government in the White Paper are conscious of the need for giving confidence to business men by monetary stability and budgetary equilibrium. They appear to be unconscious of the still greater need of giving confidence to the men and women of the country that there will be continuing demand for their services . . . (postscript, p. 273). Apart from a simple assertion of the political priority of employment, this dismissal of the confidence issue seems to be grounded on two arguments. First, it is based on an assumption that in the conduct o its own finances the state is all powerful, and does not have to worry about the 'confidence' of potential purchasers of its debt. 'The State in matters of finance is in a different position from any private citizen or association of private citizens; it is able to control money in place of being controlled by it' (para. 198). Second, and more fundamental to Beveridge's overall stance, was the view that the 'confidence' of private investors could not be allowed to determine the level of private investment.

Private Investment

In the White Paper, the foreign balance and private investment are coupled together as being both very liable to fluctuate and equally difficult to control (paras. 45–8). In his comments on the White Paper, Beveridge strongly attacked this parallel, arguing that the two components of demand were not at all on the same footing:

> the demand from other countries for British goods and services, is beyond the control of the British Government, . . . But investment at home is beyond control of the British Government only so long as the British Government chooses not to control it. Treating the foreign balance and private investment on the same footing is equivalent to treating British industry as if it were a sovereign independent state, to be persuaded, influenced, appealed to, and bargained with by the British state. (p. 261)

In fact, the White Paper does envisage enlisting the co-operation of private investment in planning investment ahead, and the use of low interest rates to promote investment. In addition, it envisages the possibility not of direct tax changes but of tax credits as a way of influencing investment (paras. 59, 61, 72).

Given these proposals, it was not unreasonable for Beveridge to argue that the White Paper, 'when critically examined, is seen to propose no serious attack on the instability of private investment' (p. 262). His own proposals envisage the creation of a National Investment Board for the purpose of controlling investment, and in particular for stabilizing private investment. The powers of this Board are not discussed in any detail. To a considerable extent, what is said parallels the White Paper. The Board is seen as attempting to enlist the co-operation of private enterprise in planning private investment; it will be able to influence such investment by the provision of finance on favourable terms; taxation policy will also be used to this end. None of these proposals is taken into any further detail. The proposal that is decisively different from that of the White Paper is where Beveridge argues that, 'if the private owners of business undertakings . . . fail, with all the help of the state and in an expanding economy, to stabilise the process of investment, the private owners canot for long be left in their ownership' (para. 300).

Meade rightly points out that, *considered as a device*, the threat of nationalization does not look very plausible as a way of stabilizing private investment: 'It is not impossible to imagine conditions in which such a method of control might be expected to cause disturbances in the behaviour of the private business men.'[10]

But, clearly, Beveridge does not see such a threat as just one device among many for stabilizing private investment. The whole thrust of the book is that, while employment policy *per se* does not require the socialization of industry, it does require the effective subordination of private investment to public policy. As Tress noted at the time, Meade's approach was one that saw Beveridge's book through Keynesian eyes, and hence saw the problem as stabilizing private investment.[11] But for Beveridge it is clear that what is envisaged is a squeeze on private investment from a very large sustained programme of public investment.

Here it is necessary to bring in a wider theme which differentiates the White Paper's approach from Beveridge's. The White Paper is concerned with two time horizons. First, there is the period of transition from war to peace, a period likely to be characterized by shortages rather than unemployment (Foreword to the White Paper). Only after this period will unemployment threaten, as it did following the boom after the First World War. While Beveridge too foresaw a postwar boom, he envisaged beyond that a 'period for which this Report is designed—Britain immediately after the transition from war to peace—there are common objectives calling for planning, as decisively as the war calls for planning today. This period is envisaged here as a reconstruction period—twenty years more or less . . .' (para. 202; see also para. 35).

In this period, apart from dealing with the balance of payments, 'We have to destroy the giant evils of Want, Disease, Squalor and Ignorance, which are a scandal and a danger. We have to raise our output per head by improving our mechanical equipment . . . There are common objectives, which, when stated, command general assent; all of them involve planned rather than unplanned outlay' (para. 202). For this reason, Beveridge dismissed pursuing demand expansion by tax reduction ('Route III' in his book) and focused on increases of public outlay.

It is in the context of the chosen route of 'a long-term programme of planned outlay' that Beveridge's proposals on private investment (and, indeed, other areas) should be understood. In this context, while policy will be aimed partly at stabilizing private investment, it will centrally be aimed at a major programme of 'additional investment to reconstruct Britain's out-of-date capital equipment' (para. 209). Beveridge emphasizes that such a re-equipment would be aimed at raising productivity to the American level, and that this is likely to take at least twenty years (para. 212).

Public Investment

Beveridge's most oft-quoted comment on the White Paper is that it was 'a public works policy, not a policy of full employment' (p. 262). Beveridge sees this as the other side of the coin of the White Paper's lack of belief (or indeed intent) in stabilizing private investment. If private investment continues to fluctuate, public investment will be used to offset these fluctuations. But, as Beveridge pointed out, most public investment could not be turned on and off like a tap because much of it is tied to the general economic activity of the nation, is complementary to private investment, and is not postponable. (The White Paper also makes this point, though the emphasis is rather different: para. 62.) At best, perhaps some housing and road construction could be used for ironing out fluctuations in demand, but Beveridge thought this would be adequate only for offsetting movements in overseas demand, leaving little scope for offsetting changes in private domestic investment (para. 254). Hence, 'the reason for desiring a large extension of the public sector of business investment is not in order to de-stabilise it to fit the instabilities of private investment, but in order to make steady development on a long-term programme possible over a larger field' (254).

This argument is coupled with the view that past evidence suggests that simply stabilizing private investment would be inadequate to deal with unemployment, as suggested by the 10 per cent unemployment in Britain even in the boom year of 1937. The appropriate aim of policy is therefore to raise the average level of private investment as well as public investment. This is a central function of the National Investment Board, with its objectives of 'intelligent state planning and particular schemes of industrial re-organisation' (para. 255).

The Regulation of Consumption Expenditure

In addition to schemes for the regulation of public investment levels, the White Paper put a great deal of emphasis on the control of consumption expenditure. 'The ideal to be aimed at is some corrective influence which would come into play automatically—on the analogy of a thermostatic control—in accordance with rules determined in advance and well understood by the public' (para. 68). For this purpose, the government favoured the use of variations in social insurance contributions as a way of increasing or decreasing the level of take-home pay of workers and the profits of employers. (The White Paper also mentioned the possibility of variations in general taxation to influence consumer expenditure, but came down in favour of at most deferred credits, as 'direct variation, apart from its other disadvantages, would come into operation more slowly than an effective policy demands': para. 72.) The national insurance scheme was seen as central and novel enough to justify an appendix to the White Paper devoted to it.[12]

In his discussion of the White Paper, Beveridge attacked this device, fundamentally because it fails to 'proceed on the basis of planning for continuous steady expansion rather than on the basis of mitigating fluctuations' (p. 263). Commenting on these criticisms, Meade says that Beveridge's rejection of them is based on his perfectionism. 'Clearly if the unemployment percentage is never going to vary at all, the whole idea of thermostatic control becomes pointless.' Given the preconceptions and knowledge at the time, Beveridge certainly must have appeared very optimistic in his belief that 3 per cent unemployment was possible (see further below). But Meade's criticism to some extent misses the point. Beveridge's aim was not to iron out fluctuations so much as to permanently raise the level of demand so that there would always be 'more vacant jobs than unemployed men' (para. 4). For Beveridge, the most serious weakness of the White Paper is its failure to offer 'a long-term programme of expanding consumption demand, social and private, which should lead to maintaining investment'. In such a context, thermostatic devices are bound to appear something of a side issue.

The Target Level of Unemployment

On reading Meade's comparison of the White Paper and Beveridge's book, Jewkes summarized Meade's arguments as showing that Beveridge's proposals were often less clear than those of the White Paper, and that 'Beveridge, in setting the task of reducing average unemployment to 3 per cent, is aiming at the impossible'.[13] The White Paper itself had made a point of avoiding any target figure, but the choice of an illustrative 8 per cent in the appendix on national insurance contributions is indicative of where its authors thought the future might lie.

It is easy to say with hindsight that Beveridge was right and the White Paper wrong.[14] Unemployment throughout the 1950s and 1960s averaged close to 2 per cent. But what is of interest is how far the divergence of opinion in 1944 on the future was based on different views about what was to be the general thrust of employment policy. Here one has to be careful. It is important to note that neither Beveridge nor the White Paper envisaged the boom in private investment and foreign trade which was to drive the growth of the next thirty years.[15] But what is of concern here is whether, irrespective of the *source* of increased demand, a labour market with under 3 per cent unemployment was a plausible proposition.

Meade articulated the kind of reasons that lay behind the unargued pessimism of the White Paper.[16] He stressed that Beveridge's target implied not only the complete abolition of interwar mass unemployment, but also the halving of the rate between 1883 and 1913. Meade argued that, in comparison with that period, the extension of social welfare (and its likely further extension after the Second World War) had substantially increased the rigidity of the labour market—by effects both on the incentives of the unemployed and on the policies of trade unions.

Second, Meade argued, the ageing of the labour force would reduce its flexibility. Against this, Meade saw some offset from labour exchanges and the organization of the labour market. 'But when all is said and done, it remains extremely sanguine to set as an absolute target a halving of the level of unemployment before the First World War.'[17]

The same conclusion is drawn by Meade from his discussion of Beveridge's 3 per cent. One per cent of this is for seasonal unemployment, 1 per cent for frictional unemployment, and 1 per cent for fluctuations in international trade. Meade asserts the unreality of these figures, pointing out that it assumed no domestic demand deficiency, and hence perfect forecasting, to make national expenditure fit national income. But of course, what Meade ruled out was a general condition of excess demand, which is not only what actually happened, but what Beveridge assumed in his target of 'more vacant jobs than unemployed men'.

As Meade rightly points out, Beveridge not only assumes no structural unemployment, but assumes that this is possible without controls over industry and employment as used in the Second World War. In retrospect, also, this was clearly mistaken, what neither Beveridge nor Meade could foresee was that the general pressure of demand would be such that persistent structural unemployment would be offset by extraordinarily low levels of general unemployment, yielding an average even lower than Beveridge's 3 per cent.

Where Beveridge seems perspicacious is in foreseeing that, if the general boost to demand was strong enough, this would enable the seasonal and frictional unemployment levels to be greatly reduced below those assumed by the White Paper. On the source of this boost to demand he was wrong, but in its broad effects on the labour market he was broadly right. (It should be noted in passing that Beveridge, like the authors of the White Paper, was aware of the inflationary dangers of a tight labour market. But, important as this issue became in later decades, it was not central to either his book or the White Paper. Meade, however, put considerable stress on this aspect in his commentary.[18])

Beveridge and the Role of the State

Beveridge's book several times asserts that his is a policy that is compatible with both private enterprise and the socialization of industry. 'The basic proposals of this report are neither socialism nor an alternative to socialism; they are required and will work under capitalism and under socialism alike, and whether the sector of industry conducted by private enterprise is large or is small' (para. 300; see also paras. 17–18, 46–7, 270, 272). Such a view fits very well with the dominant economists' view of 'Keynesian' demand management in the 1940s, namely, that it was a device that, in Robbins's words, would 'fall outside the categories of pre-war political controversy'.[19]

Clearly, such a 'technocratic' view of demand management fitted well with the mood of the mid-1940s, in which pre-war politics was seen to have obstructed the application of rational policy measures. But Beveridge's repeated claims that neither the public/private status of industry nor the general role of the state was at issue in his proposals is difficult to reconcile with the specifics of many of these proposals.

As already noted, Beveridge's policies unambiguously aimed at reducing the scope for private investment decisions, so that, even if private receipts from investment were maintained, this would be accompanied by a major attenuation of the capacities of private investors to determine the pattern and timing of investment. Here, of course, Beveridge was able to point to passages in Keynes's *General Theory* which asserted that 'the duty of ordering the current volume of investment cannot safely be left in private hands' (p. 271).[20]

But the view of the authors of the White Paper (and probably of the Keynes of 1944[21]) was that a device such as variations in national insurance contributions was important precisely because it *did* avoid any state regulation of the pattern of output—a view common to the 'liberal/socialist' Meade and the more conservative Jewkes.[22] By contrast, Beveridge's hostility to this device was not just that it was only mitigatory (or too small in scope to be effective[23]), but that in the reconstruction period consumer choice should not be allowed full rein in determining the pattern of output.

Beveridge begins his discussion of this issue by noting the view that 'the placing of adequate purchasing power in the hands of the citizens, so that they will spend more, should be the main instrument of a full employment policy. It is argued that in this way full employment might be achieved with a minimum of state interference and planning' (para. 257). To this view he puts a number of possible objections.

First, he argues that demand must be directed in the short run to those industries and areas where unemployment currently exists, and only in the long run to where consumers' untrammelled desires would lead (para. 258). Second, increased consumers' expenditure might not be directed to those forms of consumption 'which were socially most desirable; it might go to luxuries rather than necessities .. under the pressure of salesmanship' (para. 258). Third, 'there are many essential services which individuals either cannot get for themselves or can get only at excessive cost, compared with the cost of collective provision' (para. 259).

Commenting on those arguments, Beveridge suggests that it is the third which is of central importance. On the first objection, control of industrial location and occupational mobility should be sufficient to match consumer demand and political labour supply. On the second, freedom to spend should arguably be seen as an essential liberty, and high-pressure salesmanship dealt with directly. But the third objection is seen as pre-eminent and untouched by such criticism. 'There are vital things needing to be done to raise the standard of health and happiness in Britain which can only be done by common action, which in a community where democracy is so well established as in Britain will be secured in accord with the wishes of the citizens by the democratically controlled state' (para. 261).

Beveridge contrasts what had happened in the thirty years before 1939 with what could be done after the war. In that earlier period, the spending power of the community rose by nearly one-third. 'That rise no doubt conferred great benefits but it did not abolish Want. It left the giant evils of Squalor, Disease and Ignorance still strongly entrenched' (para. 260).

In this way Beveridge returns to the great themes of his Report of 1942.[24] But what is more important in the current context is that both the work of 1942 and that of 1944 saw full employment as having a clear objective. In 1942 full employment was seen as one of the preconditions of the success of the social welfare proposals. In 1944 the assertion of the inescapability of a government commitment to full employment was coupled with the view that 'employment is not an end in itself: it is a means to an end. The end is the abolition of great social evils . . .' (para. 244). Hence, 'In the total outlay directed to maintain full employment, priority is required for a minimum for all citizens of housing, health, education and nutrition, and a minimum of investment to raise the standard of life of future generations' (para. 262).[25]

The Beveridge and White Paper Projects

A document as lengthy and complex as Beveridge's book cannot be reduced to a single, simple theme. Equally, the White Paper, while much briefer, was plainly the product of much complex discussion and debate, and very much the product of many hands.[26] Nevertheless, enough has been said above to suggest a substantial divergence between the approaches of the two documents.

For Beveridge, full employment was to be based on a long-term programme of planned outlay with clear social priorities for the foreseeable future. Within this private investment might continue, albeit reduced in scope from prewar levels, but this would be conditional on its subordination of this programme of planned outlay. The context of full-employment policy was unambiguously to be a major extension of the role of the state and the public sector, in the sense of both increased expenditure by state bodies and increased state determination of the decisions of non-state bodies.

While Beveridge paid repeated lip-service to the compatibility of his programme with public or private ownership, the White Paper was more truly compatible with a continued dominance of the private sector. In the White Paper full employment has no stated purpose or objective. Hence there is no role for a programme of *planned* outlay. Instead, what is needed is a series of devices, compatible with the continued dominance of consumer choice, to reduce fluctuations in demand. As a project for such devices, Beveridge's plan is not especially persuasive—which is essentially Meade's argument.

This contrast should not be overdone. As hardly needs restating, both Beveridge and the White Paper mark a watershed in public perceptions of public policy, and both were strongly marked by some similar intellectual and political background.[27]

But beyond this rather broad point, their differences appear more striking than their similarities. This is not just a case of Beveridge being the more full-heartedly 'Keynesian', as Winch suggests.[28] Whatever precise meaning is given to that adjective, it seems unhelpful in differentiating between Beveridge and the White Paper. Harris, in her biography of Beveridge, has seen Beveridge as having moved from an earlier advocacy of a planned economy to a Keynesian analysis of full employment by 1944.[29] While undoubtedly Beveridge's position had changed, and the Keynesian element is apparent, this view, like Winch's, seems one-sided. Beveridge's book is a long way not only from the conservative Keynesian of Jewkes, but also from the 'liberal–socialist' version of Meade. To both, the idea that full employment has a clear purpose, and that that purpose is to be expressed via a long-term and substantial extension of the state's role in determining expenditure priorities, seems wholly alien. In so far as that term can be given a clear meaning, it does seem at variance with what can sensibly be called 'Keynesianism' (at least in the 1940s), even if Beveridge was able to quote some of Keynes's more radical sounding remarks in support of his arguments.

Conclusions

This chapter has not discussed the issue raised by Jewkes of whether the Beveridge project was compatible with a continuation of freedom. Equally, it has not dealt with Meade's view that Beveridge's plan did not give any convincing account of how private investment was to be effectively subordinated to the programme of planned outlay.

Neither of these issues was of course put to the test. Neither Beveridge nor the White Paper's proposal bears much relation to the full-employment period of the 1950s and 1960s. Certainly Beveridge proved right in that public investment could not be readily used as a counter-cyclical device—a lesson that was learnt only slowly after the war.[30] But on the Budget, neither Beveridge nor the White Paper foresaw that the perennial issue of the acceptability of budget deficits would be rendered essentially beside the point (for thirty years, anyway), as high private investment nullified any need for large public deficits.[31] The Budget did become the centrepiece of economic management, but around a trend rate of low unemployment which was largely the consequence of non-budgetary sources—basically, private investment. This meant that, as favoured by the White Paper and opposed by Beveridge, consumption was the item in national expenditure that formed the immediate aim of most macroeconomic management. But equally, Beveridge's social priorities were asserted, albeit largely in a separate sphere of policy, labelled the 'creation of the welfare state'. This in turn facilitated macromanagement in the buoyant world of the 1950s and 1960s, by providing a large basis of public expenditure and hence taxation to manipulate the level of demand.

The issue that remains unclear is how far the suppression of Beveridge's 'Great Evils' was indeed predicated on full employment, as he asserted in 1942. Or is it more appropriate to see suppression of those evils as simultaneously the means and the outcome of full employment as argued in 1944? With the re-emergence of mass unemployment and a 'crisis of the welfare state', this seems to be the issue which in the long run is most significant in the discussions of the 1940s.

Notes

1 While Beveridge used (and defined) the term 'full employment', the White Paper committed itself only to 'high and stable' employment. As Beveridge and Jewkes noted, the expression 'full employment' does occur twice in the White Paper, having 'slipped in'. J. Jewkes, 'A Defence of the White Paper on Employment Policy 1944', Ch. 3 of *A Return to Free Market Economies?* (London 1978), fn. 18.

2 This is apparent from the Economic Section material produced to defend the White Paper, PRO T230/69 and also by Keynes, e.g. Keynes to Barlow, 15 June 1944, in J. M. Keynes, *Activities 1940–46: Shaping the Post War World: Employment and Commodities* (Collected Writings, 27; London 1980).

3 D. Winch, *Economics and Policy: A Historical Survey* (London 1972), 263.

4 This has been a recurrent theme of Jewkes's writing: PRO T230/69, J. Jewkes, *Notes on Report of Private Conference on Full Employment held by Nuffield College, 9–10 December 1944* (29 January 1945); J. Jewkes, 'Second Thoughts on the British White Paper on Employment Policy', in National Bureau of Economic Research, *Economic Research and the Development of Economic Science and Public Policy* (New York,

1946), 117–8; J. Jewkes, *Ordeal by Planning* (London, 1948), 48–73; J. Jewkes, 'A Defence', 45.

[5] PRO T230/69, J. Meade, *Sir William Beveridge's 'Full Employment in a Free Society' and the White Paper on Employment Policy* (1 December 1944).

[6] Issues emphasized in both Beveridge and the White Paper, but ignored here, are industrial location policy, the mobility of labour, and the whole international dimension.

[7] J. M. Keynes, *The General Theory of Employment, Interest and Money*, (Collected Writings, 7; London, 1973). p. 376.

[8] For the 1930s, see R. Middleton, 'The Treasury in the 1930s: Political and Administrative Constraints to the Acceptance of the "New" Economics', *Oxford Economic Papers*, 34 (1982), 48–77. For the 1940s, see Chapter 3 above.

[9] J. M. Buchanan and R. Wagner, *Democracy in Deficit: The Political Legacy of Lord Keynes* (London, 1977).

[10] Meade, *Sir William Beveridge*, para. 30.

[11] PRO 230/69, R. C. Tress, *Sir William Beveridge on Full Employment in a Free Society, the Government on Employment Policy, and Mr Meade on Both* (8 December 1944).

[12] At Sir John Anderson's insistence: PRO PREM/4/96/6.

[13] PRO CAB124/831, Jewkes to Brook (6 December 1944).

[14] Ironically, 3 per cent became the official government target in 1951—see Chapter 7 below.

[15] R. C. O. Matthews, 'Why Has Britain Had Full Employment Since the War?', *Economic Journal*, 78 (1968), 555–69.

[16] Meade was of course a (if not *the*) major contributor to the White Paper.

[17] Meade, *Sir William Beveridge*, para. 15.

[18] Ibid., paras. 20, 77–82. On this issue see R. Jones, *Wages and Employment Policy, 1936–86* (London, 1987).

[19] PRO CAB123/48; cited in J. Tomlinson, *British Macroeconomic Policy since 1940* (London, 1985).

[20] Keynes, *General Theory*, 320.

[21] Keynes, *1940–46: Shaping the Post War World*, 308–13, 317–34.

[22] Jewkes, *Notes on Report*.

[23] Beveridge, *Full Employment*, 263.

[24] *Social Insurance and Allied Services*, Cmd. 6404 (London, 1942). See also Beveridge's statement in PRO T230/69. Private conference to consider Draft of Part IV of *Full Employment in a Free Society*, Nuffield College (18/19 March 1944).

[25] Beveridge criticized the White Paper for restricting the use of essential expenditure to the transition period (pp. 269–70, fn.).

[26] See Chapter 3 above.

[27] P. Addison, *The Road to 1945* (London, 1977).

[28] Winch, *Economics and Policy*, 263.

[29] J. Harris, *William Beveridge: A Biography* (Oxford, 1977). It is probably right to say that Beveridge had no stable and clear economic views, unlike in the area of social welfare, where he had a clear and longstanding project: A. Cutler, K. Williams, and J. Williams, *Keynes, Beveridge and Beyond* (London, 1986).

[30] J. C. R. Dow, *The Management of the British Economy 1945–60* (Cambridge 1965), Ch. VII; see also Chapter 6 below.

[31] This theme is returned to in Chapter 9 below.

5

The Search for Policy Devices: Fulfilling the Commitment of the 1944 White Paper on Employment Policy

The 1944 White Paper on Employment Policy marks a turning point in the history of economic management, perhaps without precedent.[1] But, as was recognized in the paper itself, the commitment to a 'high and stable' level of employment was not supported by specific proposals for legislation.[2] There was discussion of particular policy instruments, but much of this was extremely tentative and equivocal. As discussed in Chapter 3, in addition to its well-known ambiguity over budgetary policy,[3] the paper contained discussions of variation in national insurance contributions, of controlling public and private investment, and of possibilities in the separate areas of industrial location and foreign trade. But these areas, and others mentioned only in passing, defined a series of issues for future exploration, not a clear programme for translating commitment into action. This chapter examines what became of these explorations, and of the attempt to put flesh on the bones of that commitment.

The White Paper's tentativeness did not disguise a lack of seriousness about employment, under either the wartime Coalition or the Labour government that followed. Rather, it reflected the common view of the time that government was in this area moving into uncharted seas, where both caution and experimentation were required.[4] The commitment of the White Paper was taken very seriously because of the general view that a slump in the future was extremely likely, once the abnormal backlog of demand from the war was exhausted. This expectation of a slump provides a key component of the context in which discussion of employment policy proceeded in the 1940s. While such an expectation was most acute up to 1947, it never entirely disappeared until the mid-1950s. The expected origins of the slump shifted over the years, from a worry about a general decline in domestic demand to fears either of an American recession (especially in 1949 and later, in 1953[5]) or relating to Britain's capacity physically to procure or purchase a full-employment level of imports (continually from 1947).[6] The fear of a slump contrasted, of course, with the actuality of excess demand, and this contrast hindered the developed of anti-slump measures (see below).

The publication of the White Paper in May 1944 marked no sharp break in discussion of the issues involved. In the Economic Section of the Cabinet Office, where the first draft of the White Paper originated, much discussion in the summer of 1944 focused on the organizational structure needed for implementing employment policy. By May 1945 a note of a meeting at the Treasury suggested: 'It has not been possible to make much progress with this work while the war with Germany has been at its height, but the time has come when we ought to review the administrative arrangements to make sure that the Government machine will be equal to its new tasks.'[7] In fact, the administrative change needed to pursue this new task was remarkably small. This derived from the context of policy in its early years, both intellectual and economic.

First, it cannot be stressed enough that for many officials, especially in the Economic Section, the economic planning that figured so much in the rhetoric of the period was conceived largely as national income planning, not as the sort of detailed planning of particular sectors of the economy which the phrase was and is taken to imply. This position was well summed up by a Memorandum by the Economic Section circulated in late 1945.[8] Entitled 'Economic Planning', this stressed that 'The first and foremost purpose is to consider not so much the position of particular industries or regions, but rather the overall position of the economy as a whole . . . planning should take place in terms of the broad categories of demand upon the community's resources.'

The memo went on to point out that the machinery necessary for negative controls would be totally inadequate when 'inflation gives place to deflation'. But, of course, in 1945 the obvious administrative development was to link planning for full employment with measures to cope with a situation of excess demand. When Gilbert wrote to Bridges in the spring of 1945 on the question of Employment Policy, he could argue that: 'For some years it is likely that the policy will involve keeping the brake on with varying degrees of pressure, on both capital and consumer expenditure. I see no difficulty about that, it is in harmony with all our past training and experience, and the constitution of the machinery of Government is well fitted for the exercise of negative controls.'[9]

Up until 1947, most work on employment policy took place through sub-panels of the Investment Working Party (IWP), whose main purpose was the control and planning of investment. In these early years considerable work was put in on employment policy, co-ordinated by these panels, and drawing particularly on discussion in the Economic Section, the CSO and the new Employment Division of the Treasury. But employment policy was always a subordinate part of the Working Party's overall work. The work became less administratively focused after the reorganization of responsibilities of 1947, and the IWP's successor, the Investment Programme Committee, took only an occasional interest in the issue.

Of course, policy issues do not fit neatly into committee structures, and discussion of employment policy periodically took place in other official and ministerial committees. Nevertheless, the general point is clear. While the CSO and the Economic Section could be seen as the 'Economic General Staff' called for by the 1944 White Paper, their wartime origins lay elsewhere than in employment policy, and employment, while important, was never their overriding concern. Most Economic Section work can be seen as part of the planning machinery of the time, though the Section also had rather unclear channels into the Treasury, prior to the reorganization of 1947.[10]

Although it would be unfair to exaggerate the point (especially given the late date), the following Minute of a meeting of the Committee on control of Investment illustrates something of the position of preparations for the slump in the planning machinery. At the meeting on 11 July 1949, Bretherton stressed the lack of departmental plans for projects on the shelf, which was explained by the fact that 'most Government departments were far too busy restricting the level of investment at the present time to work out any detailed plans for the expansion of investment'.[11]

Nevertheless, especially before 1947, work was done on employment policy, albeit mainly by a small group of officials, and this normally took as its starting-point the discussions in the 1944

White Paper. A major element in these discussions had been the possibility of using public investment for employment policy. However, limits to this had been foreseen, both because the urgent needs that such investment was aimed to satisfy made postponement difficult, and because central government lacked the powers to make local authorities and public utility bodies accelerate such projects (para. 62). Overall, the pessimism on what could be done to regulate the level of private investment (see below) was offset by measured optimism on the side of public investment: 'the Government believe that they can influence public capital expenditure to an extent which will be of material value for the purpose of maintaining employment' (para. 62).

As the White Paper made clear, central government was itself responsible for only a tiny proportion of investment, so that in this area it would have to work through other bodies. Many of the public utilities referred to in 1944 were of course nationalized under the Labour government, so that policy on public investment quite quickly resolved itself into two very distinct areas: local authorities, and the socialized industries (as they were then commonly called).

Public Investment—Local Authorities

In attempting to organize a policy on employment in the area of investment by local authorities, central government was following the main lines of employment policy of the interwar period. Public works investment in areas such as housing, schools, and roads had of course been major planks in the proposals of critics of government policy in the 1920s and 1930s, and much of the history of interwar economic policy has been written around such proposals.

Whatever may be concluded about the general macroeconomic stance of interwar governments as represented through their public works policies, some movement was apparent on the administrative side. Between 1920 and 1931 the Unemployment Grants Committee (UGC) provided grants to loçal authorities for unemployment relief works.[12]

In 1930 the Public Works Facilities Act had speeded up the process of land purchase by local authorities envisaging its use for such works. Unlike the UGC, this provision remained, albeit in abeyance, after 1931. In 1938 the Ministry of Health had asked local authorities for their programmes of intended investment over the next five years, though this process had not gone very far because of war preparations.

Such programmes were the starting-point for the proposals in the 1944 White Paper, which saw them being 'assembled by an appropriate co-ordinating body under Ministers and adjusted, upward and downward, in the light of the latest information on the prospective employment situation' (para. 63). Such investment would be controlled through the mechanism of finance, the sanctioning of loans and grants being regulated in accordance with employment policy. The authors of the White Paper had also plainly learned one of the lessons of the interwar debates—the slowness with which most public investment could be started. Hence they stressed the need 'to seek means of reducing the time-lag which ordinarily intervenes between a decision to undertake public capital expenditure and the actual start of the work' (para. 64).

In this area, then, there was something for policy to be built upon in the 1940s. From the spring of 1945, Bridges began to organize the relevent administrative action. An Investment Working Party was set up, but even before this became operational a circular had been sent by the Ministry of Health in October 1945 to local authorities, and similar circulars were addressed to the public utilities. These followed Chester's memo, with which he had followed up his earlier pressing of Robbins in the Economic Section to get work going in this area.[13] The memorandum and the circular that followed it both focused on obtaining information from local authorities about their investment intentions (over a three-year period), rather than on issues of the control of that investment, though it was widely recognized that for the foreseeable future governments would continue to restrict local authority and all other investment levels.

When the Investment Working Party structure was organized in 1946, sub-panels were set up on public investment and private investment. The remit of those panels was to 'see what preparatory work ought to be done to ensure that the machinery for planning investment and using it as an instrument of employment policy is in working order when the time comes'.[14] From the beginning, a dominating theme of discussions in these panels was that of building up a 'shelf' or 'reserve' of works to be brought into use when unemployment necessitated it. However, it proved very difficult right from the start to get local authorities to do much on this front. The reason at its broadest was the difficulty of getting bodies that were intensely preoccupied with frenetic reconstruction plans, and subject to severe control in doing this by central government, to devote serious attention to developing projects for the contingency of future unemployment.

The Investment Working Party's Report of 1946 recommended measures aimed at overcoming this reluctance. These focused on making it easier for local authorities to borrow to purchase land for future schemes—something that the wartime controls on financing had specifically excluded. Ministers agreed to this proposal, and local authorities were encouraged by circular to buy land and carry through the preliminary work for a 'shelf' of projects up to the point where they could be put out to tender.[15]

But by late 1947 the whole context of discussion was shifting, as fears of a future depression retreated and the contradiction between severely controlling local authorities' current investment and encouraging them to prepare future investment plans seemed more acute. It was decided not to ask local authorities for their investment plans for the financial year beginning in 1948, as these plans now bore so little relation to any likely outcome.[16] There was considerable discussion about how far the 'shelf' remained relevant. The Lord President and the Lord Chancellor were said to be keen that this measure should be continued, though the Treasury voiced doubts about the potential inflationary impact of the enhanced borrowing for substantial land purchases by local authorities that this would involve. It was also pointed out that, *pro tem*, a 'shelf' existed simply as a result of the excess demand for investment.[17]

Another question that was increasingly raised against the 'shelf' approach to 1947 and 1948 was whether it was to prepare such projects without considering balance of payments implications. In early 1948, fears of what would happen if no Marshall Aid was forthcoming heightened the belief that unemployment might occur, but this unemployment was now seen as coming directly or indirectly through the balance of payments. As the chief planning officer, Plowden, noted in March 1948, since the Investment Working Party had reported a year earlier the situation had changed—the worry now was 'not a general falling off of trade, but lack of hard currencies to purchase vital imports'.[18]

Such concerns led to a continued concern for the building up of a reserve of works, but with a proviso that such schemes should be assessed with respect to their import content as well as their employment-creating potential. A Report by the Investment Programmes Committee in March 1948 (the body that succeeded the Investment Working Party) argued that, 'where the amount of imported material per thousand tons of contract value does not exceed one ton of steel, together with half a standard of soft wood or the equivalent in other materials, the project should be included within the reserve of works'.[19]

These changes of focus at the level of central government seem to have had little effect on what happened at the level of local authorities, where little was done to build up such a reserve. The Report of the Committee on the Control of Investment in 1949 outlined three reasons for this lack of response, two basic and one more passing:

> First, the local authorities have been preoccupied with the urgent work immediately before them; second, local authorities are not interested in buying land and spending money on plans for works which will be put in hand only at an uncertain date when a slump is definitely diagnosed; and, third, it was not possible to give a positive lead to Local Authorities right from the start because of the need to discourage purchases in the period before the Appointed Day under the Town and Country Planning Act, 1947.[20]

In response to this failure, the Committee suggested that the whole idea of a 'reserve' or 'shelf' be given up, and that instead the emphasis should be on keeping a proportion of the backing of local authority projects in readiness for rapid implementation. In this way it was hoped to avoid having two categories of local authority project, one currently programmed and one located in a vague never-never land of contingent mass unemployment. The idea of a specific reserve would be relevant only if and when the backlog of local authority projects substantially diminished. This was coupled with a proposal that local authorities should have a general right to establish reserve funds, from which the relevant minister would be able to sanction expenditure on advance preparations for works schemes.

The financial aspect of the reluctance of local authorities to prepare works for the future was met, at least in part, by a scheme for the Treasury to defray the additional costs of preparing such plans. But the idea of a specific 'reserve of works', attacked in 1949, remained a concern of policy until at least the middle 1950s.[21]

In addition to these financial changes, central government moved some way to increase the speed at which public works projects could be implemented. The 1946 Acquisition of Land (Authorization Procedure) Act extended the scope and further speeded up the process of land acquisition by local authorities. This did something to reduce the lags in such plans, which were estimated at two to three years in 1947 (before the 1946 Act became effective).

The discussion so far has focused on the administrative proposals emanating from the central authorities, and on the lack of response at the level of local authorities. This focus seems largely appropriate on the basis of the public records, where the picture is mainly one of the centre trying to stimulate an unwilling periphery into action. However, within some central departments, also, enthusiasm was lacking—as the above quotation from Bretherton demonstrates. Nor was the general notion of making local authority public works the centrepiece of employment policy taken for granted, even by those directly concerned with that policy. Periodically, scepticism on this issue is apparent.

For example, in the early days of policy discussion in 1946, an Australian economist, David Bensusan-Butt, had stressed that the provision of skilled workers for building and civil engineering projects might remain a bottleneck even in a slump, if the decline in demand originated in the export sector. This was a point frequently made in the discussions prior to the 1944 White Paper.[22] Another point, also reminiscent of the Government's 'notorious' White Paper of 1929 and the pre-1944 debates, was the importance of labour mobility to the effectiveness of such projects.[23]

In 1948 a Central Economic Planning Staff paper on the 'Reserve of Works' was criticized for its focus on local authorities:

> The type of works envisaged in the memo are mainly the types of public works which had been tried out in the 1930s and which had comparatively little effect on the total employment situation. Works of this kind could not be expected to employ more than 200,000 people directly and these

unskilled men fit for heavy labour. The schemes in the main made no real provision for employing women and skilled men.

It was also argued that the emphasis should be on private industry, which could react more quickly to emergencies and covered an enormously wider range of employment'.[24]

Such scepticism did not however switch the focus of policy in this area, and this for a number of reasons. First, immediately at the end of the war such scepticism was overriden by the fear that depression was close and preparatory measures of some sort, vital. For example, Alec Cairncross, economic adviser to the Board of Trade, invoked the lessons of the interwar period: 'In 1930 I believe that Tom Johnston found it very hard to get the local authorities to think anything up.'[25]

While such fears of a 'general' depression had abated by 1947 and 1948, there were still very strong worries about unemployment arising from an inability to finance a full-employment level of imports, and/or a fear of the effects of an American recession.

In the face of such worries, public works appeared as one area where government could translate its employment commitment into actual measures fairly readily, at least in comparison with the proferred alternatives. And while such optimism was undoubtedly undermined by the failure of local authorities to respond, this could be seen as reflecting the abnormal pressures on those authorities arising from the scope of reconstruction, rather than as a fundamental obstacle to using them as instruments of employment policy. Moreover, public works, with all their difficulties, looked more plausible as measures of control of a slump than either investment by the nationalized industries or private investment (see below).

Public Investment—Socialized Industries

The 1944 White Paper had not envisaged the major extension of the public sector that took place under the Labour government of 1945. Some of those charged with the development of employment policy saw this extension as enhancing the possibility of controlling the level of investment, on the assumption that expansion of the public sector would enhance government powers. While the issue never came to be tested, the optimism here, like that over public works, was slowly eroded in the late 1940s.

The Investment Working Party established a separate sub-panel to look at the issue of employment policy and the socialized industries. In December 1946 this panel produced a draft report which argued that government 'direct influence over broad investment decisions' would be increased from 29 per cent in 1938 to 50 per cent by 1947 because of the extension of nationalization.[26] The report recognized that different nationalized industries had different investment horizons and this affected their suitability for employment policy; it also noted possible bottlenecks in firms supplying to the industries. But the overall tone of the report was one that saw nationalized industries as major instruments of employment policy.

The report's Summary and Conclusions argued that 'The extension of the Government's control over industry, besides increasing the size of that control, will alter its quality because it will be possible far more than in the past to influence investment decisions which in turn affect the investment goods industries.' And 'we recommend that socialised industries be plainly instructed to frame their programmes with employment policy in mind.'

This draft report was then circulated for comment. On his copy, Gilbert made two comments which encapsulated the two main difficulties of the report's approach. On the one hand, he noted that the industries had been socialized because they were seen as providing the necessities of life, and that this essential role gave little scope for fluctuation in provision: hence, the 'difficulty that one's powers are greatest where the scope for their exercise is least'. On the other hand, Gilbert had doubts about the exercise of these powers, and he didn't like the idea that the industries should be 'plainly instructed'; 'I would rather rely on consultation and co-operation, with the power of instruction kept in the background. Even with the socialised industries, we shall not get them to do things unless we can persuade them into the frame of mind that they want to do them.'[27]

Civil servants directly concerned with the nationalized industries also responded strongly to the report's contents. The Ministry of Fuel and Power pointed out that varying investment in socialized industries to offset private investment fluctuations would pose severe problems for their own investment planning. It also noted the potential conflict between the Board's duty to pay its way and the demands of employment policy. Similar comments come from the Ministry of Supply, which stressed that the basic consideration in the running of the socialized industries was efficient and economical working.

The report had stressed that full employment would be in the interests of the socialized industries themselves by ensuring demand for their products. Yet a Ministry of Fuel and power official, while accepting the general force of such a comment, went on to say: 'I do not think however that the Central Electricity Board would, when a slump is foreseen, embark on a programme of constructing Generating Stations, costing some millions, several years in advance of the date when normal commercial practice would lead them to start on this work, in the hope that in this way they would create employment and keep up a demand for electricity.'[28] He further argued that, if the government wanted socialized industries to go against their normal commercial judgement, they would have to give clear instructions to this end and also to compensate the industries for the losses thereby incurred.

The idea of such subsidies was anathema to the Treasury, which consistently took the view that any general system of subsidy would mean only a shifting of the timing of investment from the unsubsidized to the subsidized periods. This view embraced both public and private investment, and in the former case justified the general requirement that the newly socialized industries 'pay their way'.

In the face of such comments and considerations, the report was substantially redrafted. Unlike the first draft, the second stressed that employment was not the only reason why nationalization had taken place. It suggested that the industries should have investment projects in three ranks of urgency, with the second rank providing a 'reserve of works' for deployment when a slump occurred. The changed tone of the report is indicated by the fact that it recommended 'the socialised industries should be *invited* by the Ministers concerned to draw up their investment programmes on these lines' (my italics), and asked to get less urgent categories in preparation.

This second draft report seems to have been pigeon-holed in the face of the more pressing problems of 1947. But its tone and content reflected the trend of policy discussion over the next few years. The Report of the Committee on the Control of Investment in September 1949 noted the expansion of the share of investment by the public sector, the 1948 figure being given as 43 per cent. It pointed out that this public sector domination of investment was not so striking in the case of plant and machinery, where three-quarters of investment remained in private hands with almost all the remainder being by the nationalized industries. It drew from this the conclusion that 'The socialisation of various industries has for the first time brought an appreciable part of investment in plant and machinery within the sphere of public control' (para. 27(c)). (Yet there had been licensing of plant and machinery during the war.)

But this argument was then substantially qualified. It was stressed that 'it cannot be assumed that the mere transfer of an activity from the private to the public sector automatically guarantees that control by the Central Government will be fully effective', and that 'there are practical limits to the extent to which public investment can be expanded or contracted to compensate for fluctuations in the private sector' (para. 28). These qualifications were taken even further in the general conclusion to the Report: 'in each of the separate areas of investment, not excluding the public sector, the practical difficulties of harmonising the investment activities of the separate agencies with the investment policy of the General Government become greater the more closely they are examined' (para. 44).

As the Report noted, the whole question of the relation between nationalized industries and central government was being considered by a ministerial committee. Chester outlines the main points from this enquiry, including its discussion of investment and employment policy.[29] The discussions of the enquiry followed lines broadly similar to those of the IWP Report of 1946. A new element was the Attorney-General's view that the legal powers of the government in this area were extremely limited: 'The policy of the nationalisation Acts does not seem to have been expressed in such a way as to enable the nationalised industries to be used as instruments for promoting economic results outside their own immediate field' (p. 984).

The Report of the Ministerial Enquiry observed that in any event the relationship between the industries and ministers could not sensibly be settled by the courts. Hence it was argued that the effectiveness of the government's control over the investment programmes of a socialized industry must depend on 'co-operation, friendly relations and good understanding between the Minister and the Board concerned'. It was also accepted that, if directions were to be given to a board, then this should be clearly and publicly understood, so as to provide cover for the consequences of those directions for the finance position of the industry.[30]

The Report was optimistic that the boards, if directed, would respond positively. However, it is not clear that this view was justified. Before the Committee reported, a case had arisen where the National Coal Board had objected to the speed of expansion of coke ovens wanted by the Ministry of Fuel and Power.[31] The dispute had been eventually resolved by certain ministerial undertakings, but here was a clear case of the kind of problems that could arise, and would be plainly exacerbated if the context were one of a severe depression.

In broad terms, the problem may be seen as part of the general problem of the role and status of nationalized concerns. Nationalization was carried out for many, and not necessarily compatible, motives. In addition, the public corporation format created serious ambiguities about the scope of ministerial powers. This problem never came to the fore over anti-slump measures (though it has of course rumbled on subsequently over issues of investment policy and prices). In addition to the administrative framework of nationalization, there was also the well-recognized point that much of the industries' investment was very long in gestation, and hence difficult to bring into effect quickly. This could be offset only by the expenditure of considerable resources on forward preparation (especially expensive when the technology was changing), which nationalized industries and local authorities alike would hardly favour without subsidies from an unwilling Treasury.

Overall, as with public works and the local authorities, the belief at the end of the war that the socialized industries would provide a ready instrument of employment policy had been considerably disappointed by the time the Labour government came to an end. Some of the reasons were similar in both cases: the lack of powers of central government (especially because of an unwillingness to embark on a general policy of subsidy), the compelling concern of the local authorities and nationalized industries with their own intensive current programmes, and the seeming unreality of planning for a slump when the pressing problem of the day was insatiable demand. At the centre, Meade might urge that some 'real social inconvenience and cost must be accepted' for the sake of employment policy, but for the local authorities and nationalized industries, employment policy implied a cost they were unwilling to bear in the absence of compelling signs of benefit.[32]

The problem was not of course only an administrative one. The suitability of using investment in both public works and nationalized industries to offset slumps was questioned, though for slightly different reasons. In the case of public works, the points were very much the same as those made in the public works debates of the 1920s and 1930s, and involved such issues as the types of labour to be employed and the regional distribution of the works. For the nationalized industries, the problem was in some respects new. If the industries had been nationalized to renew the basic infrastructure of the country, did not such a renewal have imperatives different from those of the short-run demands of employment policy? In sum, both public works and the expansion of the basic industries might be seen as suitable for offsetting a prolonged slump after the slump has arisen: by the end of the 1940s it was apparent they were largely unsuitable for the kind of periodic upswings and downswings which seemed to be the problem to be combated.

Private Investment

The White Paper of 1944 had seen the most serious obstacles to the maintenance of total expenditure as including the 'highly inconvenient fact' that 'those elements in total expenditure which are likely to fluctuate most—private investment and the foreign balance—happen also to be the elements which are most difficult to control' (para. 47). While Beveridge might rail against this equation, implying the 'sovereignty of private industry',[33] its implication was clearly that the major focus of stabilization policy would necessarily be elsewhere—on public investment and consumption. But as this paragraph of the White Paper implies, the problem with private investment was not just how far it could be controlled, but also whether it was likely to be a source of fluctuations; if so, knowledge of private investment would be vital to the whole conduct of employment policy.

This point was clearly recognized in the White Paper, which

stressed the need for industry to supply the government with 'exact quantitative information about current economic movements. Without this, informed control would be impossible and the central staff which it is proposed to set up would be left to grope and flounder in uncertainty' (para. 82).

Hence the approach of employment policy to private investment after 1944 followed two quite separate strands: a search for direct information, especially given the well-recognized problems of forecasting investment from other variables; and, in the words of the White Paper, the encouragement of 'privately owned enterprises to plan their own capital expenditure in conformity with a general stabilisation policy' (para. 61).

The lack of information about private sector investment was a recurrent feature of discussion in the Investment Working Party. Much discussion took place from 1945 on the manner of approach to private industry, and this was eventually taken up at the highest levels by ministerial approaches to the Federation of British Industry (FBI). The reason for this high-level approach was the political sensitivity of relations between the Labour government and private industry.[34] The difficulties of the form that this approach should take led to considerable disagreements at the official level, which also (in part) reflected this political sensitivity.

This dispute arose initially from the view of the Economic Section and the CSO that, if the central purpose of collecting the data was to enable the government to foresee a slump, then the need for short-term information on investment intentions—up to eighteen months ahead—was paramount.[35] The Section argued that such information was vital not only for predicting a slump, but also for providing much needed experience in handling data from firms and comparing intentions with outcomes.

The Board of Trade and Ministry of Supply, by contrast, emphasized the problem of asking firms about investment intentions when in fact investment was controlled by government anyway. The exercise 'would be tantamount to asking them to guess the extent to which they thought the government was likely to grant them licences. This could only have the effect of discrediting the whole approach to industry for information about their investment programmes'.[36] They argued that information on short-term investment could best be derived from the issue of building licences and the programmes of machinery manufacturing firms. Approaches to investing firms should be restricted to their long-term forecasts for five years or so, at least until pent-up investment demand had disappeared.

This dispute rumbled on for some months, and was settled eventually largely on the Board of Trade's and Ministry of Supply's terms. The consequence of this settlement was that a lack of information continued to concern those officials concerned with employment policy.[37] This was partly owing to the very cause of the friction—the government controls over investment. In 1947 the existing building and raw material controls were supplemented by a centralized scrutiny of all investment projects over £500,000. Such controls obviously yielded considerable information to central government.

The Report of the Committee on Control of Investment noted that the consultation of private industry with government had 'developed against a background of statutory controls', and accepted that 'the disposition to consult with Departments in this way may diminish as the efficacy of those controls diminishes', but expressed the rather pious wish that such co-operation would continue once controls ceased (para. 38).

Of course, the late 1940s did see a large expansion of economic statistics, in line with the growth of national income measurement and forecasting. But the kind of data that the CSO and Economic Section were looking for—data of investment intentions which could be used for predicting a slump—were never forthcoming in the years immediately after 1945.

Parallel to, and (as noted) sometimes intertwined with, the issue of investment information was that of the control of private investment. The 1944 White Paper's general scepticism on the plausibility of this was justified by what followed.

The general view at the central government level was compounded of an acceptance that private investment was central to overall stabilization, and doubt over what could be done to control it. In the early discussions on employment policy in 1945 both these points were emphasized. The Treasury argued that private industry could not be allowed to go its own way on investment or this would put an unbearable strain on the balancing role of public bodies. Gilbert saw the need to extend the work begun on the public sector to the private sector, but 'I must confess myself at a considerable loss in knowing how to start on this.'[38] He echoed the hope of the White Paper, that the private sector like the public sector would draw up a programme of investment, and would see benefits from co-ordinating this with public investment used in an anti-cyclical manner.

Little came of this, as private industry felt unable to give any clear programmes in the absence of government guidelines as to the broader economic environment in which such investment might take place. While the government went some way in sketching its broad views of the future in the Economic Surveys, these were neither detailed nor accurate enough for the private sector. In this impasse it would be absurd to 'blame' either side, as of course the period was one of frequent crises and inescapable policy changes which precluded the kind of forecasting that was deemed desirable.[39]

Certain measures were taken in this period which, though not initiated for employment reasons, gave government at least a bit more leverage over private investment when the period of direct controls was over. One of these was the introduction in 1945 of initial investment allowances for companies to set against taxation. This was used as a way of encouraging investment in the late 1940s, especially in plant and machinery. How far such measures would be of use in a major slump was never of course tested, but scepticism on this seems appropriate.[40]

There was some discussion of the general use of tax measures to regulate the level of private investment in these years, but the Inland Revenue, who were consulted by the Treasury, were unenthusiatic. They stressed the need to maintain the revenue, and the difficulty of 'taking back' concessions once given.[41] Here, indeed, there was little evidence of a shift from long-existing departmental concerns.

Another measure cited as a potential help against recession was Section 2 of the Borrowing (Control and Guarantee) Act 1946, which empowered the government to guarantee loans up to a total of £50 million in any one year for the purpose of facilitating the reconstruction or development of an industry or part of an industry.[42] This was not designed for anti-recessionary purposes, and because it could not be used to aid individual firms directly it was likely to be slow-acting, requiring the use of *ad hoc* bodies to organize the disbursing of money.

Finally, there was the belief that, as the banks and other

financial institutions had broadly co-operated in the control of advances to industry, this process could be put into reverse in a slump. But again, some scepticism may be appropriate about the effectiveness of such measures. It would seem, in the famous phrase, like pushing on a piece of string to get industry to accept loans for investment in a recession.

In 1951, Atkinson of the Economic Section drew up a Memorandum on *Full Employment and Control of Investment*[43] which gave an optimistic view of the potential for government stimulation of private investment via tax incentives and financial assistance. Apart from the unresolvable issue of the appropriateness of such optimism, it is worth noting that these mechanisms had very little to do with policy initiatives taken since the White Paper for employment purposes, which were largely unsuccessful.

As with local authorities and socialized industries, central government faced a severe administrative and (one might say) psychological problem in persuading private firms to prepare for a slump when government was at the same time preventing them from investing as they would wish. In the case of private investment, this problem was compounded by the political problems of a reforming Labour government dealing with a private enterprise sector which saw itself as under threat. Co-operation was close in some areas, but tensions were sometimes considerable, as, for example, when negotiations between the Board of Trade and the FBI reached a deadlock in 1948 over the scope of enquiry into firms' investment intentions.[44]

Some of the discussion of economic planning under the Labour government had made plain the negative character of that planning and the surprising ignorance of the private sector on which it was based. But these points would have made the stimulation of private investment in a recession a difficult task if indeed the expected slump had come in 1947 or 1948.

The Variations of National Insurance Contributions

The clearest specific proposal in the 1944 White Paper was for a scheme of variation in national insurance contributions according to the level of unemployment. An appendix to the White Paper outlined such a scheme, and gave a numerical example centred on an 8 per cent unemployment rate.

The inclusion of this proposal in the White Paper followed considerable controversy. Meade and the Economic Section pressed the case strongly for such a scheme, as an automatic way of compensating for variations in demand originating elsewhere. The Treasury was unenthusiastic. This was partly because they foresaw difficulties in deciding on the 'normal' level of unemployment in the immediate postwar period—and this difficulty was accepted in the White Paper (para. 71). More fundamentally, Hopkins argued, there might be circumstances when such a scheme was not appropriate; 'but also, one cannot in practice expect successive governments in the long term necessarily to act always in the face of public pressure, on grounds of wisdom rather than of political expediency'. Thus, part of the Treasury objection to such a scheme was that, like the Budget generally, once the acceptability of deficits was conceded, political pressure would make them more common than events justified. This kind of pressure, Hopkins foresaw, might lead to 'the new and vast social insurance fund falling into those conditions of bankruptcy which were experienced in the unemployment fund fourteen years ago'.[45]

A further doubt, linked to general Treasury arguments, was whether such a scheme might not be used in appropriate circumstances, that is, where the unemployment level did not reflect a general lack of demand but 'for example, . . . arose from a loss of export trade which entailed a serious derangement of external payments, or . . . followed upon a phase in which costs and prices had increased sharply'.[46]

The Economic Section attempted to refute such arguments. On the first issue, they stressed that it was an automatic system, and, as Meade put it in later discussions, 'The automatic link is for this reason, in my view, the main Treasury safeguard against a continuing imbalance of the fund.' The Economic Section also suggested that to allay Treasury fears the national insurance fund might be separated into two, an ordinary and an employment stabilization fund, with the latter 'taxing' and 'subsidizing' the former as appropriate.[47]

On the issue of the appropriateness of such a measure, the Economic Section focused on its relation to structural unemployment. They argued that this objection was not decisive against such a scheme because structural unemployment was not a separate problem from general unemployment. A problem would arise only if major structural changes were to be combined with very low unemployment—an unlikely combination, they asserted.[48]

In 1944, with Keynes's enthusiastic support for 'one of the few absolutely concrete suggestions for stabilising employment', the Economic Section's view largely prevailed. Henderson eventually said he would accept the Annex to the White Paper as long as the problem of the scheme's applicability in immediate postwar circumstances was stressed.[49] Sir John Anderson, the Chancellor of the Exchequer, decided to include this annex, 'as the scheme is both novel and highly debatable'. He also stressed that the 8 per cent unemployment figure used there was purely illustrative and not a target level, which would tend to undermine the intentions of the government in this area.[50]

Consideration of this issue after the publication of the White Paper arose not from the general discussions of employment policy initiated by the Treasury, but from the Labour government's reform of social security. Discussions leading to the major statute in this area, the National Insurance Act 1946, raised the issue of whether it should include such a scheme for variations in contribution rates.

Meade argued strongly that, if such a scheme were to be operative by 1947, when he (and many others) felt a slump might occur, then it would have to be included in the 1946 Act. He consulted the Ministry of National Insurance on such a scheme and was told that, though it would involve difficult administrative problems, as far as could be seen it would not be an administrative impossibility. Meade also made the political point that, as such a scheme was the most concrete proposal in the White Paper, 'It would be a matter for serious adverse comment if the National Insurance Bill introduced by this Government failed to make any mention of this radical proposal which was agreed upon by all parties in the Coalition government.'[51]

The Treasury was opposed both to the scheme and to its inclusion in the National Insurance Bill, but this seems to have been overborne by strong ministerial support for inclusion, especially from Morrison.[52] However, the powers granted in the Bill were drafted in the widest possible terms, and specifically did not set up a sliding scale. This was rejected by Dalton on the grounds of lack of data, but also because it would involve the government in committing itself to a figure for the 'normal' level of unemployment. This was seen as undesirable because, if a high figure were set, it might be criticized as undermining government's commitment to low unemployment or, if not criticized, might engender complacency in government policy if unemployment did not exceed this target level. Alternatively, if the figure were set too low, the fund might not balance. The Treasury also successfully insisted that the scheme be deferred until 'normal' conditions returned.

Meade argued strongly for an automatic basis, because it would mean that the scheme would work quickly, avoiding any time-consuming discussions of whether the rise in unemployment was temporary or permanent. In addition, he reasserted the point that automaticity would guard against the political pressures feared by the Treasury. At the same time, he accepted that, while the principle of automaticity was vital, attaching figures to this would be premature until conditions became more settled.[53]

The outcome of all this was something of a Pyrrhic victory for the proponents of this method of dealing with unemployment. The National Insurance Act 1946 did include powers for the minister to vary contributions, but these powers were very broadly drafted, and the extent and timing of the variations would be at the discretion of the government, judging each occasion on its merits. This ran against the fundamental basis of the scheme, that it be rapid and automatic.

The proposal was not formally buried. In 1948 the UK government told the United Nations that the proposal was 'still on the table'.[54] In 1950, in discussions on the UN experts' report on measures against unemployment, the government spelt out the basis of its lack of enthusiasm for automatic compensatory mechanisms of any kind, including the national insurance fund. First was the objection made by Henderson in 1944 that such schemes 'could hardly distinguish between the causes of unemployment and might therefore be misapplied'. In addition, such automaticity threatened the tradition of ministerial discretion in determining taxation and the budget.[55]

These objectives seem to have proved decisive in stopping any such scheme going beyond the powers of the 1946 Act, even after 1950, when the government's acceptance of a 3 per cent unemployment target removed one of the previous objections to such a scheme. The Economic Section was still discussing the details of such schemes in 1950, but this seems to have had little link to policy formation.[56]

Overall, the defeat of such schemes was a victory for Treasury views. Meade attempted to counter one of the major planks of Treasury opposition by stressing that the automatic aspect would guard against fears of weak-willed politicians refusing to raise contributions when the level of unemployment demanded it. But this only led the Treasury to emphasize the problems of automaticity, both on the grounds that the average unemployment rate was too inexact a measure of unemployment conditions to trigger a change automatically in the level of overall demand, and because it reduced budgetary discretion. The potent fears of a future slump in 1946 gave the scheme's proponents a scent of success, but thereafter the scheme was permanently pigeon-holed.

Conclusions

In 1949 the Control of Investment Committee concluded its examination of possible policies in this area by arguing as follows:

> The results are disappointing. Each of the separate instruments we have examined has turned out, in closer inspection, to be a good deal less effective than might have been expected, and in each of the separate areas of investment, not excluding the public sector, the practical difficulties of harmonising the investment activities of the separate agencies with the investment policy of the Central Government become greater the more closely they are examined. (para. 44)

This paragraph could stand as a general epitaph on the attempt to develop the means of action on unemployment suggested in the 1944 White Paper.

Of course, the pressure to find a plausible solution to unemployment diminished through the late 1940s, as the threat of such unemployment was perceived to diminish. This is illustrated by the surprising lack of debate around the 1951 decision by the government to adopt a 3 per cent unemployment target—precisely the figure derided as unrealistic when put forward by Beveridge six years earlier.[57]

Nevertheless, it would be wrong to suggest that fear of unemployment had disappeared entirely by 1950. Indeed, the adoption of the 3 per cent target in 1951 was in part an attempt to get the Americans similarly to commit themselves, and thus reduce the threat of a recession originating in the United States. Rather, the declining belief in the measures adumbrated in 1944 was paralleled by a rising faith in the efficacy of 'general financial and budgetary policy' as a weapon against future unemployment.

The general debates on budgetary policy in the late 1940s cannot be dealt with here, but one point should be stressed.[58] The adoption of budgetary policy as central to economic management in peacetime can plausibly be dated to the emergency budget of November 1947. The context of this was of course excess demand in the economy, requiring a policy of disinflation rather than demand expansion. So budgetary policy was not adopted at a time when its efficacy as a weapon against unemployment was to be put to the issue. Indeed, one can say that it was precisely the fact that the issue was the scale of budget surpluses, not budget deficits, that overcame the still strong opposition to the use of the budget as a macroeconomic weapon. In the late 1940s, the threat of unbalanced budgets hardly appeared imminent.

So the kind of macroeconomic management that began to emerge in the late 1940s, and was to dominate the 1950s and 1960s, owed little to the devices suggested in 1944. It focused on budgetary and, to a lesser extent, monetary policy, within an overall framework of buoyant private expenditure and budget surpluses. The fears and constraints on budget deficits had not been directly confronted because there was no need. The mass unemployment strongly feared, especially down to 1947, was postponed indefinitely—in the event, for over thirty years.

Notes

1 See Introduction, above.

2 On 'high and stable' versus 'full' employment, see Ch. 4, n. 1.

3 E.g., D. Winch, *Economics and Policy: A Historical Survey* (London, 1972), 281.

4 PRO T161/1200/S52842/02, Report of Speech by Sir John Anderson, Chancellor of the Exchequer, to Municipal Treasurers (21 June 1945).

5 PRO CAB 134/22, EPC(49)64, *Economic Policy in a Recession*, Memo originating with the Economic Section, later circulated by the Chancellor, Cripps.

6 E.g., PRO T161/1370/S53555/011/2, Sir Bentley Gilbert, 2nd Secretary to the Treasury, to E. Bamford (19 January 1948).

[7] PRO T230/70, Note of meeting on employment policy (May 1945; no author).

[8] PRO CAB 134/186, *Official Steering Committee on Economic Development*, Memoranda.

[9] PRO T161/1297/S53555/3, Memo Sir Bernard Gilbert to Sir Edward Bridges (20 March 1945). Bridges was Secretary to the Cabinet.

[10] The fear of inflation was sometimes (especially in the Treasury) linked to unemployment by the suggestion that high inflation would cause unemployment. But the standard view was that summarized by Robert Hall when head of the Economic Section in 1948: 'The disadvantaes of continually rising prices are obvious: they are unfair to those with fixed incomes and (unless exchange rates move) they diminish exports and attract imports, thus requiring the sale of reserves or foreign borrowing. In extreme cases of inflation the economic system collapses': PRO T171/394, *Note on Inflation*. On the issue of policy on inflation in this period, see R. Jones, *Wages and Employment Policy, 1936–86* (London, 1987).

[11] PRO T229/169, *Committee on Control of Investment*, Minutes (11 July 1949).

[12] PRO T161/1297/S53555/06. F. L. Edwards, *Local Authorities as an Instrument of the Full Employment Policy*, (April 1946). Edwards was in the Ministry of Health. For official attitudes to public works between the wars see R. Middleton, *Towards the Managed Economy: Keynes, the Treasury, and the Fiscal Policy Debate of the 1930s* (London, 1985), Ch. 8.

[13] PRO T161/1200/S5284/02, D. N. Chester, *Capital Expenditure by Local Authorities* (May 1945). Chester was a member, and Robbins the head, of the Economic Section until succeeded by Meade later in 1945.

[14] PRO T161/1251/S53555/04, *Investment Working Party*, 1st joint meeting (4 March 1946).

[51] PRO T161/1296/S53555/2, *Report on Preparation of Investment Projects by Public Authority* (January 1946).

[16] PRO CAB 134/437, *Investment Programmes Committee*, Minutes (16 December 1947).

[17] PRO T161/1370/S53555/011/2, *Investment Working Party*, Memos on Investment by Public Authorities.

[18] Ibid., note by Chairman (12 March 1948).

[19] PRO T161/1370/S53555/011/2.

[20] PRO T229/237, *Report of the Committee on Control of Investment*, para. 31.

[21] PRO CAB 134/890, *Working Group on Employment Policy* (1952); CAB 134/891, *Sub-Group on Reserve of Works* (1953/4).

[22] PRO T161/1296/S53555/2, D. M. Butt, *A Note on Employment Policy* (April 1946); PRO T161/1168/S52099, *General Papers Leading up to the White Paper on Employment Policy*.

[23] The Government's response to the Liberal's *We Can Conquer Unemployment* was *Memoranda on Certain Proposals Relating to Unemployment*, Cmd. 3331, PP (1929); PRO T161/1296/SSS333/2, *A Note on Employment Policy* (April 1946).

[24] PRO CAB 134/210, *Economic Planning Board*, Discussion of Central Economic Planning Staff Memorandum on 'Reserve of Works' (1 April 1948).

[25] PRO T161/1370/S53555/011/1, A. Cairncross to P. Vinter (7 November 1946). Cairncross was at the Board of Trade. Vinter, along with P. D. Proctor, were the senior Treasury civil servants solely concerned with employment policy. But see PRO CAB 124/617, *Ministry of War Transport. The Preparation of Post-War Projects*, for difficulties in this area as early as 1944.

[26] PRO T161/1297/S53555/06, *Draft Report on Preparation of Investment Projects in the Public Sector* (October 1946).

[27] PRO T161/1297/S53555/012.

[28] Ibid., P. D. Proctor to P. Vinter (7 June 1947).

[29] D. N. Chester, *The Nationalisation of British Industry 1945–51* (London, 1975), 980–90.

[30] Ibid., p. 989.

[31] Ibid., p. 982.

[32] PRO T161/1251/S53555/04, J. Meade, Head of Economic Section, to B. Gilbert (16 September 1946).

[33] W. H. Beveridge, *Full Employment in a Free Society* (London, 1944), Postscript, 261.

[34] PRO CAB 134/438, *Investment Programmes Committee*, Minutes (9 November 1948).

[35] PRO T161/1297/S53555/06, *Investment Working Party; Joint Working Parties on Public and Private Investment.*

[36] PRO T161/1251/S53555/07, ibid., Memorandum (9 October 1946).

[37] PRO T161/1297/S53555/3, P. Vinter to P. D. Proctor (17 April 1947).

[38] PRO T161/1200/S52742/03, B. Gilbert to P. D. Proctor (27 June 1945).

[39] A. Cairncross, *Years of Recovery: British Economic Policy 1945–51* (London, 1985).

[40] On their effects in the 1940s and 1950s, see J. C. R. Dow, *The Management of the British Economy 1945–60* (Cambridge, 1965), 204–9.

[41] PRO T229/169. *Committee on Control of Investment*, Minutes (18 July 1949).

[42] PRO T229/237, *Report of Committee on Control of Investment*, para. 41.

[43] PRO T229/323, *Full Employment and the Control of Investment*, by F. Atkinson.

[44] PRO CAB 134/438, *Investment Programmes Committee*, Minutes (9 November 1948).

[45] PRO T161/1168/S52099/01, R. Hopkins, Permanent Secretary to the Treasury, to Chancellor of the Exchequer (29 April 1944).

[46] Ibid., H. Henderson (no date); see also various papers in T161/1168/S52098.

[47] PRO T230/105, J. Meade to Lord President (January 1946); T230/72, Economic Section comments on Social Security Variations (23 October 1943).

[48] Ibid., T230/72.

[49] PRO T161/1168/S52099/01, *Papers on Variations in Social Insurance Contributions, 1944.*

[50] Ibid.

[51] PRO T230/105. Memorandum from J. Meade to Lord President (5 December 1945).

[52] Ibid.

[53] Ibid.

[54] PRO T230/230. *Draft Reply of the UK Government to the UN's Questions on Employment Policy.*

[55] Ibid., *Experts Report on Full Employment Measures*, by J. V. Licence (8 September 1950). Licence was a member of the Economic Section.

[56] PRO T230/202, *Papers on National Insurance Scheme—Working Papers 1949–54.*

[57] E.g., PRO T230/230, *Committee on National and International Measures for Full Employment*, Minutes (5 February 1951).

[58] A. Booth, 'The "Keynesian Revolution" in Economic Policy-making, *Economic History Review*, 38 (1983), 103–23; N. Rollings, 'The "Keynesian Revolution" in Economic Policy-making: A Comment', *Economic History Review*, 38 (1985), 95–100; A. Booth, 'The "Keynesian Revolution" in Economic Policy-Making: A Reply', *Economic History Review*, 38 (1985), 101–6.

6

The Keynesian Revolution of 1947?

This chapter takes up the themes of Chapter 3 in looking at the crucial year for economic policy, 1947. In that earlier chapter some of the broad questions of interpretation of changes in economic policy were used to introduce a detailed discussion of the 1944 White Paper. These questions of interpretation related particularly to the role of changes in economic theory, as opposed to other elements, in bringing about shifts in perceptions of the possibilities of economic policy. The earlier chapter linked these issues to the production of a document; this chapter links them to an undoubtedly significant change in the actual conduct of policy which took place in 1947. The argument here is not about the fact of such changes but about their significance; in particular, how far they justify the concept of a 'Keynesian revolution' in that year.[1]

Some of the earliest dicussions of economic policy change in mid-twentieth-century Britain proclaimed a startling belief in the capacity of ideas to change policy. In this view, Keynes's theoretical ideas, once resistance was overcome, opened the way to the banishment of unemployment which in the 1950s and 1960s appeared a permanent achievement. M. Stewart epitomizes this view:

> Mass unemployment was brought to an end by the Second World War. It has never returned. Despite dire predictions in 1945, Britain has now enjoyed full employment for more than thirty years. Such a tremendous transformation might be expected to have many causes. But in fact the evidence points to one cause above all others: the publication in 1936 of a book called *The General Theory of Employment, Interest and Money* by John Maynard Keynes.[2]

Such views fitted in with Keynes's own oft-repeated notion, which saw the obstacle to the enactment of his policy proposals largely in the stupidity of administrators, businessmen, and politicians.

Other more detailed historical work later qualified this picture of slow intellectual enlightment, but still focused its main attention on the theoretical perceptions of actors in the policy-making process. Winch, for example, while well aware of the difficulties of focusing on ideas, does use a framework that fits with that view of how policy changes.[3] This emphasis on the economic theory adhered to by actors in the policy-making process has recently been criticized from a number of rather diverse angles.[4] Such work is unified by its focus on elements other than economic ideas in understanding the undeniable changes in economic policy between the early 1930s and the postwar years.

At the same time, the dating of the Keynesian revolution has tended to be pushed forward. Howson and Winch saw a process of substantial education of officials in the 1930s, laying strong foundations of ideas which could be taken to their logical conclusion during the Second World War.[5] Peden's work, however, has thrown considerable doubt on the idea of any such substantial theoretical shift in the Treasury before the Second World War.[6] More recently, Booth in particular has emphasized 1947 as a crucial date when the final victory of Keynesian policies was achieved, albeit this is seen as a culmination of a long process, not a sudden revelation.[7]

Before looking at the events of 1947 and their significance, some problems of definition need to be aired.

Problems of Definition

These variations in interpretation and timing of the Keynesian revolution raised the issue of the definition of that revolution; how far is there an agreed definition, or do the disagreements cited above simply reflect the different definitions adhere to by different authors?

Winch offers a very broad definition of the Keynesian revolution: 'not simply a matter of employment policy, still less of counter-cyclical action. It is best thought of as a rational approach to the problems of economic management in general, based on the conscious assessment of the outcomes of alternative lines of action.'[8]

Such a broad definition clearly has its merits; in particular, it opens up the question of why national economic management took such different forms in different countries, and thus avoids equating Keynesianism with the peculiar features of much of

British postwar policy.[9] However, as a definition it raises problems because it seems difficult to see why anyone should be opposed to 'a rational approach to the problems of economic management'. The rub is of course the word 'rational', about which we can say the same as Hirst says of 'reasonable': 'the one piece of knowledge I feel any certainty about is that men can say *anything* and consider it reasonable.'[10] If everyone agreed on what was rational in economic management, then the need for a Keynesian revolution would seem to disappear.

Booth[11] follows Sayers[12] in defining the Keynesian revolution in terms of the acceptance of budgetary arithmetic as a weighty—but not the only—element in macroeconomic policy. On this basis, it is apparent why Sayers (and others) have seen 1941 as a crucial turning-point—that is the year when for the first time the Budget was used to try to manage total expenditure in the economy.

Clearly, this use of the Budget was new and can plausibly be said to be revolutionary against the traditional view of the role of government Budgets. However, there would seem something inherently paradoxical in seeing a policy of minimizing budget deficits or maximizing surpluses, aimed against inflation, as a triumph for a line of reasoning that had its origins and *raison d'être* in a concern for unemployment. Moreover, a focus on the use of the budget *per se* as a weapon of economic management makes it difficult to see what all the fuss was about. Why should this revolution be so long resisted, so hard fought, if it involved only a new way of deploying the Budget? Against this view, I want to argue that the Keynesian revolution should be defined in terms of an attempt to legitimize deficits as a device for use when the level of aggregate demand required stimulation.[13]

Focusing on deficits means arguing that budgetary policy is strongly assymetrical— that budget surpluses pose few of the controversial issues posed by deficits. Deficits are a problem at two different levels. For the department concerned with the control of public expenditure, in Britain the Treasury, they raise the spectre of political pressures for the expansion of such expenditure without proponents of such expansion facing the constraint of also contemplating higher taxes to finance the expenditure. As Middleton argues for the Treasury in the 1930s, 'the precept of the balanced budget acted as the ultimate constraint on the growth of expenditure, since it moderated and tempered the natural demands of politicians and sectional interests for new expenditure, and provided a "neutral" framework within which competing demands were met.'[14] This political dimension, as opposed to any theoretical consideration, seems to have been crucial to the Treasury opposition to 'Keynesian' proposals.

Linked but separable from this concern was the effect of budget deficits on confidence. Given that the chosen way of financing budget deficits is through private financial markets, the maintenance of confidence is vital to successful budgetary policies: 'to a large extent the success of fiscal operations in a capitalist society depends on their being favourably received by financial markets.'[15]

This constraint then exists at two different points: the perceptions by government of what private markets will stand, and the markets' own perceptions, though presumably the former will tend to shift broadly in line with the latter. Of course, market perception of acceptable government fiscal stance will alter, and at times (as in the early 1980s) financial confidence may make absorption of government debt ready and uncontroversial. But the general point is that the importance of such market perceptions means that budgetary policy can never be simply subordinated to the demands of macro-management.

Such a role for financial markets is of course conditional on financing deficits by sales in such markets. This in fact has been the path commonly advocated by Keynesians, though a case can be made that, given a Keynesian underemployment equilibrium, financing via money creation (i.e., borrowing from the central bank) is just as sensible an option.[16] This, for example, was the kind of scheme proposed by Meade in the 1930s, for the Unemployment Assistance Board to be financed directly by the Bank of England, with the notes issued being cancelled when the Board's receipts exceeded its expenditures.[17] For whatever reasons, this approach was not adopted, and the precepts of 'sound finance' meant that finance could never simply be 'functional', in the sense of Lerner.[18]

Focusing on the legitimation of budget deficits as central to the Keynesian revolution is not without its difficulties. First of all, Keynes's own view on budgets was complex. In his *Can Lloyd George Do It?* and *The Means to Prosperity* Keynes went out of his way to argue the limited effects of his policy proposals for public deficits.[19] But this itself suggests his awareness of the sensitivity of the issue. In the debates leading up to the 1944 Employment White Paper, Keynes espoused the cause of a separate capital budget in order to maintain balance in the current Budget.[20] The grounds for this opposition were partly theoretical, some arguably muddled, perceptions of the likely effects of short-run changes in income on personal consumption.[21] But more important, in the current context Keynes feared that, 'If taxes were lowered, it would be hard to raise them again. If deficits were proposed, the proposal would provoke opposition and perhaps weaken confidence—a possibility that was recognised in the *General Theory*.'[22]

Keynes's view in this area, as in others, seems to have been that the stupidity of non-economists had to be reckoned with, not simply ignored. Hence his comment on Lerner's discussion of budget deficits: 'he shows that, in fact, this does not mean an infinite increase in the national debt . . . His argument is impeccable. But, heaven help anyone who tries to put it across the plain man at this stage of the evolution of our ideas.'[23]

On this basis, it seems reasonable to argue that Keynes's theoretical position led him to advocate budget deficits, while his well-known desire to offer practical proposals led him to advocate devices for avoiding the appearance of such deficits. In this way one of the inherent difficulties of 'Keynesianism' was present in the master's own policy prescriptions.

Equally, the stress on the difficulties in legitimating budget deficits does not imply that such deficits could not be produced as a result of compelling political circumstances—as in 1919/20.[24] This episode, however, hardly represents a break with basic policy, as deflation was pursued as soon as it was seen to be politically possible, and no rationale other than the perceived threat of Bolshevism underlay the policy. Equally, the loan-financed rearmament expenditure of the late 1930s represented very little departure from the traditional treatment of war expenditure as legitimately financed by borrowing.[25]

To focus on budget deficits is not to attempt to reduce the complexity of economic policy-making to one simple formula. But it is to try to clarify the extent and significance of the changes in economic policy that undoubtedly occurred between the early 1930s and the end of the 1940s, in a manner consistent with the fundamental thrust of Keynesianism, the provision of a solution to unemployment.

What Happened in 1947?

In 1950 Robert Hall, head of the Economic Section, wrote: 'The last Government adopted in 1947 and 1948 a revolution in British practice, when they took responsibility for maintaining full employment, but avoiding inflation. It is the best argument in favour of this policy that the revolution passed almost unnoticed.'[26] This view of the importance of the policy changes of that year has recently been endorsed by Booth and Cairncross. These authors focus on the budget of November 1947, Cairncross seeing it as a 'turning-point in post-war fiscal policy',[27] and Booth making the much broader claim that it was 'a major milestone, when the Treasury finally turned in peacetime and out of choice to Keynesian analysis to help control inflation'.[28]

Hence it is the basis of the Emergency Budget of November 1947 that needs to be looked at to see why such significance has been accorded to it. This Budget seems to have been founded on a combination of the short-run problem of the sterling crisis following the premature abandonment of exchange control, and the longer-term pressures for anti-inflationary measures.

The possibility of a supplementary Budget was first mooted by the chancellor of the Exchequer, Dalton, on 11 August.[29] This was in the very middle of the convertibility crisis, only six days before convertibility was suspended.[30] According to Dalton's Memorandum, the Budget's 'sole purpose, if it were decided to have one, would be to lessen the inflationary pressure by "mopping up" some purchasing power, and at the same time to illustrate the Government's intention that there shall be "equally of sacrifice" as between different sections of the community'.[31] The timing and tone of this memo supports Cairncross's view that the intention behind Dalton's proposal was as a 'public acknowledgement of danger', in a way parallel to previous use of the Bank rate, now of course ruled out by the adherence to cheap money.[32]

On the official side, the Chancellor's floating of the idea of a supplementary Budget, and the crisis that stimulated the idea, provided an ideal opportunity to press for strong anti-inflationary measures. On 1 September Robert Hall circulated to the Budget Committee a paper on *The Inflationary Pressure* which reiterated the warnings of his predecessor, Meade, about the dangers of inflation.[33] Such warnings seem to have been taken more seriously than previous ones within the Treasury. Bridges wrote to the Chancellor that the purpose of the Budget would be 'not the old-fashioned one of extra taxation to balance the budget, but a reduction of the inflationary pressure which threatens to prevent the emergency measures from achieving their objective'. He stressed that presenting the Budget in this way would maintain confidence in the value of the pound, especially in the United States, and would also help to 'sell' Marshall Aid to Congress. In his discussion of precise policies to increase the budget surplus, Bridges focused on the cost-of-living subsidies, a reduction in which, he said, was 'the pivot of the whole Autumn Budget, and indeed its one absolutely essential feature'.[34]

On his copy of the memo, Dalton wrote 'I don't accept a lot of this.' This comment has been interpreted as a sign of Dalton's resistance to 'Keynesianism',[35] but the comment seems to have been aimed not at the general framework of Bridge's proposals, but at the focus on food subsidies. A large part of the discussion of the emergency Budget was taken up with the question of subsidies, with Dalton pressing the case for cuts in expenditure on the armed forces as a superior alternative, both because of equity considerations and because the movement of manpower from military to civilian uses would increase supply in the economy, unlike the cuts in transfer payments like the cost-of-living subsidies.

In this resistance to cuts in food subsidies, Dalton was supported by Douglas Jay, who stressed the effect of such cuts on wage bargaining.[36] This was in fact the line taken by the Treasury in discussions prior to the April 1947 Budget, and indeed, cost-of-living subsidies had been introduced to reduce inflationary pressure during the war. Booth sees the attitude to cuts in food subsidies as a litmus test of 'Keynesianism', and the reversal of official views on this between the two Budgets of 1947 as a sign of their conversion to Keynesianism.[37] But the view expressed before the earlier Budget that such subsidies reduced inflationary pressure was not incompatible with an acceptance that they also increased inflation by their effects on demand. Such a view was stated by Meade, for example, who was strongly opposed to the subsidies for their distorting effects as much as their inflationary consequences.[38]

Hence it seems right to play down any change of intellectual allegiance of officials in the summer of 1947. What changed was their perception of the scale of the threat of inflation and of political possibilities. On the latter point, Gilbert seems to have been opposed to Meade's proposals for a rapid cut in the scale of subsidies because of the political repercussions, and the fear that it could lead to inescapable pressure for rises in national insurance benefits.[39] From such a viewpoint, what altered in the summer of 1947 was the political possibilities, notably the government's evident desire to stress its financial responsibility. In the event, by the time the Budget occurred, Dalton considered that internal political pressures not to cut subsidies were paramount, and the Budget abolished clothing and footwear subsidies, but simply stabilized the amount available for food subsidies. If the Cabinet had dithered less in the summer of 1947, perhaps these subsidies would have been cut, but by November the immediate crisis was past.[40]

As Booth rightly emphasizes, the Treasury officials turned to the strongly Keynesian Economic Section in the summer of 1947 because they offered techniques that seemed to provide a way of defeating inflation.[41] These officials had previously been sceptical of some of the Economic Section's policy proposals, but the techniques related to neither the principle of anti-inflation nor the analytic framework, but more to the political plausibility of the Section's proposals to deal with it. Civil servants are above all gaugers of the politically possible, and this is paramount in the shift in policies they pressed on the Chancellor between the two Budgets of 1947.

Why was the Treasury so worried about inflation? Part of the reason was the general view of inflation current at that time, of its effects on the balance of payments, with a fixed exchange rate, and its effects on those on fixed incomes.[42] But the Treasury also had a more direct concern with the likely effects of inflation on cheap money, because of the effects of higher interest rates on government debt payments. This concern was the major reason why cheap money had come about in the first place,[43] and why it had been so enthusiastically embraced by Treasury officials at the time of the National Debt Enquiry of 1945.[44].

Also, as Rollings rightly emphasizes, the pursuit of surpluses, especially via cuts in food subsidies, chimed in with the traditional Treasury concern to control public expenditure. They do not seem to have taken the view that 'public spending is only out of control if it is growing faster than government revenue':[45] rather, the Treasury concern was with the global total of public expenditure, as well as with the overall budgetary position. This is illustrated by, for example, Gilbert's response to an Economic Section Memorandum of 1949 (when the budget surplus was bigger than 1947). In this he expressed great scepticism about the value of public expenditure projections for 1955, on the grounds that one didn't need a lot of 'elaborate arithmetic' to see the problems of controlling public expenditure.[46]

The crucial issue in 1947 was not, then, the theoretical allegiances of officials. The same may be said of those of the Chancellor. But it is worth noting that, seen in the context of the time, Dalton's 'non-Keynesianism' may be seen to be exaggerated.[47] Certainly his first three Budgets betrayed little interest in the overall pressure of demand in the manner advanced by Meade and the Economic Section; his concern was to present his Budgets as a rapid return to budgetary balance in a traditional sense.[48] Nevertheless, it is worth nothing that, like most people on the left at this time, Dalton's primary concern was with the prevention of unemployment. And in this context Keynesianism appeared less relevant in the years after 1945: maintaining general employment was not an immediate problem. Thus, in his first Budget Speech in October 1945, Dalton stressed the switch from prewar deflation to the dangers of inflation.[49] In the following year he stressed that he was very concerned with unemployment in the places where it continued to exist—in the Development Areas. 'One of the first instructions which I gave when I became Chancellor of the Exchequer last July was that, as regards constructive plans for the Development Areas, the Treasury was henceforth to be no longer a curb, but a spur.'[50]

His general budgetary analysis was always that of balancing the Budget over a series of years. Within that, he focused on elements—such as changes in the level of private savings—which later became unfashionable but at the time were commonly discussed, and which can be seen as compatible with an overall income–expenditure approach to the economy.[51]

It is plain that Dalton retained a faith in microeconomic planning that was at variance with the 'liberal–socialist' emphasis on macro-planning of James Meade and other Keynesians. This difference stemmed in part from Dalton's emphasis on income redistribution, always a strong concern of his, and one for which he saw such measures as subsidies as vital.[52] But it needs to be stressed that for Dalton the system of controls that he inherited from the war seemed to be well adapted to the circumstances of inflationary pressure in the immediate postwar years, as well as fitting with his ideological predilictions. In this light, the shift towards a greater emphasis on the Budget's use against inflation can be seen as the consequence of short-term worries about the balance of payments,[53] within a general perspective which was not hostile to Keynesianism, but which regarded it as less relevant in the absence of general unemployment (see further below).

A Keynesian Revolution?

By the late 1940s, British economic policy-making had plainly altered a great deal since the 1930s. The Budget had become the central instrument of economic management, and this was undoubtedly a major change. It was a process that started in 1941 with Kingsley Wood's Budget, and was consolidated in the late 1940s by the re-integration of economic and financial policy in the Treasury. Such demand management, in an inflationary environment, required much less of a shift in the traditional concerns of the Treasury than would have been the case if the context had been one of mass unemployment. In an inflationary context, as Gilbert had written in 1945, such use of the Budget would 'involve keeping the brake on which varying degrees of pressure, on both capital and consumer expenditure. I see no difficulty about that; it is in harmony with all our past training and experience, and the constitution of the machinery of government is well fitted for the exercise of negative controls.'[54] Hence it required no intellectual conversion for the Treasury to accept this new way of pursuing its traditional concerns. Equally, with cheap money, as already noted, while this was undoubtedly a Keynesian proposal of the 1930s, it also offered the Treasury a way to a traditional goal.

Investment control has been cited as another instance of a Treasury acceptance of Keynesian remedies.[55] Here again, though, one has to have regard for asymmetries. On the one hand, Treasury agreement to control investment in order to contain inflationary pressure fitted easily into its established terms of reference. More problematic is the idea of expanding investment in time of slump. As Rollings rightly points out, after the 1944 White Paper, and with Treasury involvement, 'positive steps were taken to establish a "shelf" of projects to offset a depression and to advance the state of preparedness of this shelf as far as possible'.[56] But even here, one may question how far this involved any substantial shift in policy, given the reluctance of the Treasury to move from encouraging local authorities to establish a 'shelf' of projects, to adopting any policy of positive financial inducements to investment in a future slump. In the absence of that, the discussion of the 'shelf' appears somewhat trivial as a possible weapon against a major decline in output.

None of this is to deny that by the late 1940s the dominant discourse in economics, and among economists involved in policy formation, was a Keynesian one, in the sense of agreement with the income–expenditure framework, and with seeing the Budget as a major weapon to alter the level of aggregate expenditure.[57] Of course, this framework and its implications were much less widely known and accepted among officials or most ministers and politicians.[58] What is less certain is whether this is the crucial consideration is discussing the extent of a 'Keynesian revolution'. For, as has been suggested above, what would seem to be vital is the perception of financial markets of what is 'acceptable' in policy, and in turn the perceptions of policy-makers of what those markets will find acceptable in the context of a policy of budget deficits to counter general unemployment.

Here the issue becomes inescapably hypothetical. There was no mass unemployment in the 1940s against which to assess the response of policy-makers and financial markets, and the Budget came rapidly into surplus. All that can be offered is evidence on the position of policy-makers on policy ideas that might involve budget deficits.

First, as noted above, Dalton continually stressed his idea of the need to balance the Budget over a series of years, so that surpluses in times of inflation would earn the right to deficits if unemployment threatened.[59] What is striking in reading the budget papers is how far this gloss on current surpluses was entirely Dalton's own—there is no sense of its being the Treasury's view, and plainly it could well be intepreted by anti-Keynesians as a harmless flourish to buy political support for surpluses now.

Of course, no one can doubt the seriousness of some officials in searching to find measures to offset any future depression. But whenever such policies implied budget deficits, however disguised, substantial hostility was apparent. Meade was prolific in trying to think up schemes that tried to take account of such sensitivity to budget deficit. In his paper on *The Control of Inflation* in June 1946 he advocated a scheme which in certain respects mirrored Dalton's idea of balancing the Budget over a number of years: he advocated the creation of an Employment Stabilization Fund to be credited with the current budget surpluses, which were to be spent in bad years.

> One advantage of this system would be that it might make it possible to run a real Budget surplus next year when, from the economic point of view, a surplus is certainly required. Simply to run a surplus and to call it a surplus on a sinking fund might be unacceptable. On the other hand, to explain that a general re-organisation of financial arrangements was being undertaken in order to plan for full employment, that this special fund was being set up for this purpose, and that the economic circumstances of the time and not merely budgetary orthodoxy requires such a surplus for the time being at least, would give an entirely different flavour to the question.[60]

Unlike Dalton, Meade explicitly coupled this with maintaining an annual balance on the ordinary budget.

Despite the transparent concern with Treasury sensitivities on the issue, this proposal received a hostile response. Gilbert spoke of the proposal disparagingly: 'This is not finance, and is not economics, but is a mixture of politics and psychology.'[61]

Hostility on these grounds by Gilbert is in many ways paradoxical, for it seems that 'fiscal revolutionaries' like Meade were more sensitive to the confidence issue than fiscal conservatives like Gilbert. The evidence suggests that Meade (and Dalton) were right to perceive the need to maintain financial confidence by a City whose views on economic policy had shifted much less than that of most economists.

In 1947 City opinion pressed for Dalton's 'public acknowledgement of danger', though opinions were divided on whether the scale of that acknowledgement went far enough. *The Banker* saw the November Budget as 'a dismal failure to face the facts'.[62] *The Economist* was more encouraged, though it coupled this with a general condemnation of Dalton as 'the worst Chancellor of the Exchequer of modern times'.[63] Generally, City opinion seems to have taken a line parallel to that of the conservative wing of the Treasury—a primary concern with inflation, an emphasis on the need for big cuts in expenditure, and a focus on reductions in subsidies as crucial. There was less concern with the details of budgetary arithmetic.

> Nor need criticism rest upon abtruse and debatable calculations about the extent of the inflationary gap. The proper function of his budget, regarded from the general standpoint, was to do what all the grim speeches and exhortations and directives and cuts in rations have demonstrably failed to do—namely, to bring home unmistakably to the names of the people the real extent of the extremity, to compel or inspire in every home a crusade for personal economy and forthright effort.[64]

There was plainly a lack of financial confidence by the City in Labour's policies, especially from late 1946. This was reflected in the failure of Dalton's attempt to drive down long-term rates of interest below 3 per cent, as debt sales turned negative and the yield on consols rose to over 3 per cent. That this loss of confidence did not lead to more of a crisis is attributable to the movement into budget surplus from 1947, the fact that cheap money prevailed in other countries, and the fact that in any case external financial flows were under tight control.[65]

None of this is perhaps surprising, nor does it prove what the attitude of the City would have been if mass unemployment had come in the late 1940s, and the government had attempted to counter with fiscal policy. Direct comment on such an eventuality is absent from the financial press at this time—it was not an issue. Nevertheless, some idea of likely City response may be gauged from the *Bankers Magazine* comment in April 1947 that 'The cheap money policy derives from the thirties, and like other hangovers from that period, in particular the doctrine of the unbalanced budget, it is totally unrelated and unsuitable to existing circumstances.'[66]

It is inherently impossible to prove a negative. Nevertheless, the evidence still seems to me to be wanting that in the late 1940s there was such a fundamental shift in policy as to justify talking of a 'Keynesian revolution', in the full sense of that term.

Notes

1 I am particularly grateful for Alan Booth's helpful comments on this chapter, given his opposition to much of the argument it contains.

[2] M. Stewart, *Keynes and After* (Harmondworth, 1972), 13.

[3] D. Winch, *Economics and Policy: A Historical Survey* (London, 1972). 24–6.

[4] J. Tomlinson, 'Why Was There Never a "Keynesian Revolution" in Economic Policy?', *Economy and Society*, 10 (1981), 72–87; R. Middleton, 'The Treasury in the 1930s: Political and Administrative Constraints to Acceptance of the New Economics', *Oxford Economic Papers*, 34 (1982), 49–77; N. Rollings, 'The "Keynesian Revolution" and Economic Policy-making: A Comment', *Economic History Review*, 38 (1985), 95–100.

[5] S. Howson and D. Winch, *The Economic Advisory Council 1930–39: A Study in Economic Advice during Depression* (Cambridge, 1977), 163–4.

[6] G. Peden, 'Sir Richard Hopkins and the "Keynesian Revolution" in Employment Policy 1929–45', *Economic History Review*, 36 (1983), 281–96.

[7] A. Booth, 'The "Keynesian Revolution" in Economic Policy-making', *Economic History Review*, 36 (1983), 103–23; 'Defining a "Keynesian Revolution" ', *Economic History Review*, 37 (1984), 263–8; 'The "Keynesian Revolution" and Economic Policy-making: A Reply', *Economic History Review*, 38 (1985), 101–6.

[8] Winch, *Economics and Policy*, 293.

[9] K. Smith, 'Why Was There Never a "Keynesian Revolution" in Economic Policy? A Comment', *Economy and Society*, 11 (1982), 223–8.

[10] P. Hirst, 'The Necessity of Theory', *Economy and Society*, 8 (1979), 434.

[11] Booth, 'A Reply', 101–2.

[12] R. Sayers, '1941 – the First Keynesian Budget', in C. H. Feinstein (ed.), *The Managed Economy* (Oxford, 1983), 107–17.

[13] J. Tomlinson, *British Macroeconomic Policy since 1940* (London, 1985), Ch. 5.

[14] Middleton, 'The Treasury in the 1930s', 55.

[15] Ibid., 60.

[16] J. Buchanan and R. Wagner, *Democracy in Deficit: The Political Legacy of Lord Keynes* (London, 1977), 32–3.

[17] J. Meade, *An Introduction to Economic Analysis and Policy* (London, 1937), 51–6.

[18] A. Lerner, *Economics of Control* (London, 1944).

[19] Booth, 'A Reply', 263–4.

[20] J. Keynes, *1940–46: Shaping the Post War World: Employment and Commodities* (Collected Writings, 27; London, 1980), Chs. 4, 5.

[21] T. Wilson, 'Policy in War and Peace: The Recommendations of J. M. Keynes', in A. P. Thirlwall (ed.), *Keynes as a Policy Adviser* (London, 1982), 55–8.

[22] Ibid., 56.

[23] Keynes, *1940–46: Shaping the Post War World*, 320.

[24] Booth, 'A Reply', 264; A. Booth and S. Glynn, 'Unemployment in Interwar Britain: A Case for Relearning the Lessons of the 1930s?', *Economic History Review*, 36 (1983), 339.

[25] G. Peden, 'Keynes, the Treasury and Unemployment in the Later Nineteen Thirties', *Oxford Economic Papers*, 32 (1980), 1–18.

[26] PRO T171/400, Hall to Prime Minister, *Budget Policy* (16 March 1950).

[27] A. Cairncross, *Years of Recovery: British Economic Policy 1945–51* (London, 1985).

[28] Booth, 'The "Keynesian Revolution" ', 123.

[29] PRO T171/392, Dalton to Bridges (11 August 1947).

[30] Cairncross, *Years of Recovery*, Ch. 5.

[31] PRO T171/392, Dalton to Bridges (11 August 1947).

[32] Cairncross, *Years of Recovery*, 424.

[33] PRO T171/392, Robert Hall, *The Inflationary Pressure* (1 September 1947).

[34] PRO T171/392, Bridges to Chancellor of Exchequer (23 September 1947).

[35] Booth, 'The "Keynesian Revolution" ', 122.

[36] PRO T171/392, D. Jay to Chancellor of Exchequer, *Budget and Food Subsidies* (10 October 1947).

[37] Booth, 'The "Keynesian Revolution" ', 121–2.

[38] PRO T171/389, J. Meade, *The Inflationary Pressure* (June 1946).

[39] Compare PRO T171/389, B. Gilbert to E. Bridges (14 March 1947) with PRO T171/392 Gilbert to Chancellor (14 August 1947).

[40] Cairncross, *Years of Recovery*.

[41] Booth, 'A "Keynesian Revolution" '.

[42] PRO T171/400, Hall, *The Inflationary Pressure* (1 September 1947).

[43] Winch, *Economics and Policy*, 213.

[44] PRO T230/95, *Draft Report* by R. Hopkins (15 May 1945).

[45] Rollings, 'The "Keynesian Revolution" ', 99; Booth, 'A Reply', 102.

[46] PRO T230/151, Gilbert to Hall (10 March 1949).

[47] B. Pimlott, *Hugh Dalton* (London, 1985), 486–94.

[48] E.g. PRO T171/386, Dalton to Defence Committee (February 1946).

[49] PRO T171/371, Draft Budget Speech (22 October 1945).

[50] PRO T171/386, Draft Budget Speech (March 1946).

[51] PRO T171/371, Draft Budget Speech (22 October 1945); compare T171/403, where Gaitskell asked for investigation of a policy of selling council houses as a way of raising the level of savings.

[52] Pimlott, *Hugh Dalton*, 473.

[53] H. Dalton, *Principles of Public Finance* (London, 1954), 243–5.

[54] PRO T161/1297/S53555/3, Vinter, *Notes on the Future Role of the Investment Working Party*, citing Gilbert (5 June 1947).

[55] Rollings, 'The "Keynesian Revolution" ', 96–7.

[56] Ibid., 96.

[57] Tomlinson, *British Macroeconomic Policy*, Ch. 5.

[58] Booth, 'A Reply'; PRO T171/400, Atlee to Cripps (13 March 1950).

[59] PRO T171/390, Draft Budget Speech (March 1947).

[60] PRO T171/389, J. Meade, *The Control of Inflation* (June 1946), para. 75.

[61] PRO T171/389, Gilbert to Bridges (14 March 1947).

[62] *The Banker*, 263 (December 1947), 141.

[63] *The Economist*, 153 (22 November 1947), 832. This condemnation was very much linked to City hostility to Dalton's cheap money policy—on which see Cairncross, *Years of Recovery*, Ch. 16.

[64] *The Banker*, 263 (December 1947), 141.

[65] Cairncross, *Years of Recovery*, Ch. 16.

[66] *Bankers Magazine*, 1237 (April 1947), 275.

7

The Acceptance of Full Employment 1948–1951

In the last three years of the Labour government, economic policy was dominated by the balance of payments problem. The threat of unemployment arising from a lack of domestic demand appeared remote, and work in preparation for such an eventuality noticeably slackened. Domestic policy came to be dominated by the fear of inflation.

Towards the end of the Labour government two noticeable events (or rather, one event and one eventual non-event) occurred in employment policy. The first was the adoption of a quantitative target for employment. The other was the preparation, though not the parliamentary pursuit, of a new White Paper on full employment. This chapter discusses these two events and how they related to the development of policy on unemployment in this period.

The 3 per cent Employment Standard

In 1951 the UK government for the first time adopted a quantitative unemployment target—3 per cent.[1] This figure was precisely the same as that advanced by Beveridge in his *Full Employment in Free Society* in 1944, and widely derided in official circles at that time.[2] In the intervening years Beveridge's figure, far from being widely optimistic, had proved, if anything, conservative, with registered unemployment rarely moving above 2 per cent during 1945–50 except in the fuel crises of 1947. The adoption of the 3 per cent target figure in 1951 could be seen simply as a recognition of this unexpected but welcome buoyancy of the economy. But this would be much too simple an explanation. In 1951 there were widespread doubts about adopting a target, especially one so low. These doubts arose both from the inflationary fears that were growing in strength at this time, and from the political hostage to fortune that such a target would seem to provide.

The main reason for the eventual adoption of the target, despite these fears, arose from considerations of international policy. Central was the continual worry about the US economy—both the dangers of recession there, and doubts about the US government's willingness and capacity to offset such a recession. Such worries were a significant element in the whole discussion of the new institutions of international economy policy—which began, of course, long before the Second World War ended.

In his discussion of the British responses to the wartime US proposals for postwar policies of free trade and capital flows, R. N. Gardner notes that one important qualification to UK enthusiasm for such proposals was the employment consequences. In particular, Britain was concerned that countries should have some protection against depressions originating abroad. 'British opinion was particularly concerned with the danger of post-war depression in the US. The British response to the American challenge would have to insist that some means be found to ensure that an American slump did not occur, or, failing this, that the UK and other countries be permitted to protect themselves against the spread of the slump to their own economies.'[3]

These fears of an American slump entered strongly into the wartime discussions of Article VII of the Lend-Lease agreement, by which Britain, as a 'consideration' for Lend-Lease, was to commit itself to lift discriminating trade barriers after the war. British fears that the United States was overly concerned with the removal of trade barriers at the expense of measures of economic expansion led to an eventual version of this Article which included provision for action 'directed to the expansion, by appropriate international and domestic measures, of production, employment, and the exchange and consumption of goods, which are the material foundations of the liberty and welfare of all peoples'.

The plan put forward by Keynes for the reconstruction of the world financial system reflected this preoccupation with the importance of domestic expansion.[4] Criticisms of the alternative US White Plan expressed similar concerns.[5] In the parallel discussions on postwar trade some difference of focus was apparent. On the whole, the United States (especially the State Department) saw trade barriers as a cause of unemployment, their removal as a major step to full employment. Conversely, Britain tended to see the absence of clear policies for high employment, especially in the United States, as undermining the quest for freer trade. This was a fear relating not only to Britain but to other Commonwealth countries, notably Australia and New Zealand. Britain saw difficulties in 'selling' multilateralism to these countries unless their emphasis on domestic employment was understood and taken on board.[6]

While public opinion in Britain was moving in a leftward direction during the war, in the United States the movement was the other way. One index of this was the rejection by Congress in 1945 of the 'full-employment' commitments in the Full Employment Act of 1945.[7] In this context, negotiations on multilateralism usually recognized the interdependence of full employment and free trade but without tying themselves to specific policies on the former aspect—although, as Gardner notes, such policies would be difficult to embody in an international agreement.[8]

In the negotiations towards an International Trade Organization (ITO) in London in 1946, the difference of emphasis between Britain and the United States was again apparent. While Cripps emphasized the primacy of employment, the United States saw the charter of the ITO as having general commitments to full employment, with regard to both domestic policy and international collaboration, but commitments that were clearly subordinate to the trade proposals. The British, urged on by Australia and New Zealand, though employment should have its own clear place in the charter, and emphasized the obligation that individual countries owed to others to maintain full employment. A similar division was apparent over the extent to which a creditor country should bear the main responsibility for correcting a balance of payments position that caused employment problems elsewhere.

On both issues there was a compromise. Unresolved in 1946 was the extent to which international action on employment was possible. Britain pressed for the co-ordination of the action of the various agencies in the employment field, via the Economic and Social Council of the UN. This removed the issue to another arena, where, very slowly, the matter was to be taken up in a significant way (see below). The overall conclusion of the London conference was a series of compromises between the divergent views, with the United States continuing to emphasize the benefits of free trade except under exceptional circumstances, and Britain pressing for the acceptance by the United States that domestic employment considerations would justify import restrictions.[9]

Ultimately, after much negotiations an ITO charter was drawn up in Havana in 1947. But it was still-born, as neither country ratified it. At root, the problem was the continuing one of the US primary concern being trade discrimination, the UK's, that of protecting domestic policies of full employment. 'The result was an elaborate set of rules and counter-rules which satisfied nobody and alienated nearly everybody. They grew into such a mountain of complexity that the ITO finally collapsed of its own weight.'[10]

The debates over the ITO aptly summarize the difficulties commonly perceived in the UK as obstructing full assurance of US actions in the event of depression. In this perception, Britain in the mid-1940s was engaged in a long campaign of trying to get full employment accepted as a goal of international co-operation as well as of national policy. In a retrospective account of this campaign, Fleming said:

> this was our role at Bretton Woods, the Philadelphia Conference of the ILO, the San Francisco Conference, at the various conferences leading up to the Havana Charter of the ITO, and at many meetings of the organs of the UN and of the OEEC. Broadly speaking, we have sought by various means to encourage the adoption by governments of high standards of achievement in the field of full employment, to foster the international exchange of information regarding the techniques of domestic policy, and to ensure that international action in the field of commercial and financial policy was so adapted as to mitigate the consequences of depression in one country on the balance of payments and on internal demand in other countries.[11]

From 1947, these objectives were pursued mainly in the Economic and Social Commission of the UN (ECOSOC). Unsurprisingly, perhaps, the information exchange aspect of these objectives was most readily attained. In March 1948 the UN asked member countries to submit detailed information on how they maintained full employment and what they proposed to do in the face of a decline. In Britain a draft response was drawn up by the Economic Section; this was criticized in certain respects by the Treasury as giving too explicit commitments on what would happen in the event of a slump. In particular, the Treasury succeeded in writing out any explicit commitment to the variation of national insurance contributions.[12] Overall, however, the British reply was a serious attempt to encourage others to state their proposals in this area.

This informational aspect of British policy was institutionalized in the form of semi-annual reports on the domestic employment policies of governments. But the answers to questions on the balance of payments issues in relation to unemployment were considered less satisfactory, and this led to a further British initiative in 1949 for ECOSOC to establish an Experts' Committee on 'national and international measures to achieve full employment'. This committee included one British economist (Nickolas Kaldor), two Americans (J. M. Clark and A. Smithies), plus P. Uri from France and E. R. Walker from Australia.

The report of these experts received considerable attention in the UK and elsewhere.[13] On domestic policy its recommendations suggested that governments should:

(i) announce a full employment target;
(ii) announce a programme for directing all the relevant aspects of policy to the continuous achievement of full employment;
(iii) adopt and announce a system of automatic and compensatory measures to operate in case of failure to maintain employment;
(iv) announce its anti-inflation policies;
(v) adopt its legislative, administrative, and statistical procedures to the implementation of full employment.

In addition, the experts recommended international policies to:

(i) establish a programme, involving statistical exercises, conferences, and a permanent expert advisory commission, to eliminate the present structural disequilibrium in world trade;
(ii) stabilise the flow of international investment;
(iii) stabilise external disbursements in the face of internal fluctuations in effective demand.

For the British government, the first of these domestic policy recommendations raised a dilemma. Britain had initiated the Experts' Report and so was obliged to take it seriously, yet such targets raised difficulties both of principle and of the precise value to be adopted. The Interdepartmental Working Party set up to consider the Report spelt out these difficulties.[14] First, they made the point that such a target would not be linked to any notion of the cause of unemployment, and hence could not discriminate between cases where the unemployment was structural or frictional as opposed to arising from a shortage of demand. The Report envisaged a range of figures for the target, but the Working Party argued:

A maximum figure sufficiently high to allow some elasticity in economic policy and for accidental hazards might cause public criticism; the minimum figure might be sufficiently low to take account of a probable assumption that a level of unemployment below this minimum signifies a state of inflation requiring deflationary measures. Yet it is desirable that the UK should set its maximum and minimum as close together as possible in order to influence other Governments to commit themselves closely.[15]

They also pointed out the lack of international comparability in unemployment figures.

The general tone of the Working Party Report was scepticism about the practicability of many of the measures suggested in the Experts' Report, especially on the international aspects,[16] but there was also a strong desire to protect Britain from external depressive influences, by getting other countries to pursue policies of full employment as advocated in the Report. Fleming's report on the discussion of the Experts' Report at the Economic and Employment Commission of the UN in January 1950 followed a similar line of thought. It stressed that Britain's interest was to get other countries, and especially the United States, to accept the idea of unemployment targets, and this was seen as the most practicable of the experts' proposals.[17]

The issue of automatic compensatory measures raised a problem for the Treasury, where such measures had long been opposed as inflexible and derogating from the chancellor's powers. These points were reiterated in the Working Party's Report.[18] On the other hand, there was a desire to see the United States adopt such measures. This problem of appearing to say 'do as I say not as I do' was cleverly defended by Fleming. He argued that such measures were desirable in the United States because of the absence of controls over prices, wages, and dividends—but that in Britain such controls were themselves 'automatic stabilizers'.[19] He did not however make the rather obvious point that such controls (apart from not in fact being automatic), were much more plausible as controllers of inflationary pressures than as stimuli to expansion.

Overall, the major emphasis of UK policy following the Experts' Report was on the employment target (later called a 'standard'). While this raised domestic difficulties, it did not directly contradict a policy position of the Treasury, as did, for example, automatic stabilizers such as variations in national insurance contributions. Britain also endorsed the proposals to announce programmes for full employment. The emphasis on anti-inflationary policy was welcomed, too, though the experts were unclear on whether they endorsed a formal wages policy.

At the ministerial level, the Working Party's Report was seen as rather negative. Gaitskell (still at this time Minister of Fuel and Power) said that he '[did] not agree with the Working Party's anxieties on the technical difficulties and [thought] the real difficulty political—to persuade the other Governments to intervene by direct controls to prevent inflation, and, in the case of the US, to take the necessary action to offset the surplus in their balance of payments'. The Prime Minister's Economic Adviser, Douglas Jay, said that 'it was vital that we take a positive attitude in ECOSOC and avoid even the suspicion of detailed criticism to cold-shoulder the proposal'.[20] This was the general feeling of ministers—to offer broad support for the Experts' Report and save any detailed criticism for a later date. Hence at the ECOSOC meeting in August 1950 Britain proposed a broad acceptance of the Experts' Report, stressing the employment target, and recommending that these targets be appraised by the ILO and ECOSOC. It did not however ask governments to pursue the domestic policies advocated by the experts. Similarly, it endorsed the principles of international action advocated while not advocating the concrete schemes.

The US draft resolution called on governments to announce not an employment target, but a running series of annual targets relating to all the main indices of the state of the economy, including employment. It also disregarded the British experts' specific recommendations on domestic policies. On the international side, it avoided any reference to the topics of stabilizing international investment or preventing the spread of depressions, and confined itself to an elaboration of the long-term equilibrium at which countries should aim.

Weakened by the 'defection' of Australia (following the general election there), Britain was not strongly supported. Broadly, the final resolution coupled the US and British positions, with an endorsement of targets but also of annual reports on economic plans and prospects; governments were not committed to any particular domestic policies on employment, but were asked to define their attitude to such policies. On the international side, the idea of public investment to maintain total foreign investment and foreign holdings of surplus country's currency did not survive, but the United States did accept the need to try to prevent shortfalls in foreign investment or other recessionary factors from causing unemployment in other countries.

As Fleming noted in his report on this meeting, its resolutions plainly obliged Britain to establish a full-employment standard. But in addition, it required Britain to outline its attitude to various potential policies against recession for which 'something like a new Employment White Paper is required, stating the government's attitude towards various techniques of policy. There is every advantage in getting this out early, so as to give a lead to other countries.'[21]

In London, Bridges was appointed to chair a working party on the implementation of the ECOSOC resolution. This working party reported in December 1950. In a covering memo, Bridges stressed the novel nature of the standard, and argued for careful consideration because of its 'great political significance'.[22] The report itself reiterated some of the problems with a standard—the varying sources of unemployment, the problem of inflation. On the former, the report foresaw a reduction in structural unemployment through industrial distribution policies. On the latter, it argued that a 'slightly higher level' of unemployment than the recent figure of 2 per cent would suffice. After considering the problems of publishing a target range, the report plumped for a 4 per cent single target. This figure was based on an estimate of the likely *external* recessionary forces that the UK might face—it assumed no problem arising from domestic demands. The exercise meant guessing about the effects of a US recession, using 1937/8 as the basis. This yielded an estimate of 2 per cent extra unemployment (allowing for British counter-measures) to add to the normal 2 per cent at the seasonal peak. Hence a 4 per cent target.

Gaitskell, in proposing a standard to the Cabinet Committee, proposed 3 per cent. He accepted that the various causes of unemployment might justify a slightly higher figure, but he argued:

First, the public may not always recognise that action will start before the Standard is reached, and possible misrepresentation on this point will be less serious if the Standard errs on the low rather than the high side. Secondly, the main practical advantage of the publication of national Standards will be to encourage the United States to publish and (when the time comes) to observe a reasonable standard and for this purpose also the lower we can put our own Standard the better.

This proposal was agreed on 3 March 1951. The actual statement to ECOSOC said that the margin above 2 per cent was to allow for export decline and inflation.[23]

The announcement of the target was made in March 1951. This was the first public commitment to a target unemployment figure, though in development area policy a (higher) figure had previously been in use as a standard.[24] It became the figure around which official discussion was conducted under both the Labour government and the Conservative government which came to power later in 1951.[25]

The White Paper that Never Was

Britain's attempts to encourage US policies of full employment led via the 1950 ECOSOC resolution to the 3 per cent target. They also, as noted above, set up pressures for a full statement by Britain of her policies against any threatened slump. This was one of the factors that led to the drawing up of a new White Paper on Employment Policy in 1951, intended as a successor to the 1944 original. However, this White Paper represented the convergence of many diverse policy considerations, some quite far removed from the normal concerns of employment policy.

An initial impetus came from the need to review emergency powers in the economic field, which were due to expire with the Defence Regulations in December 1950. It was decided to take the opportunity to look further ahead and see what statutory economic powers would be needed in the long term. To this end, a committee under Gilbert was established in summer 1949 and reported in March 1950. Eventually its proposals were embodied in an 'Economic Powers Bill', which mentioned full employment as one of the purposes of the Bill but was concerned mainly with negative physical controls.[26]

Even before the Bill was drafted, however, the Cabinet had

decided that the general and wholly negative powers of the Bill were politically dangerous in the period before the 1950 general election, and the matter was put off until after that election was over.[27] The question was considered in the Lord President's Committee in April, and the argument was made that some positive role for controls was needed, specifically against deflation.

The official committee set up to consider the draft Economic Powers Bill met in the summer of 1950 and had detailed discussions of the draft. But by July 1950 the Bill had undergone its first change of title to the Economic Planning and Full Employment Bill. This arose from discussions in the Lord President's committee, where it was emphasized that what was needed was to draw attention 'to the need for economic planning in order to secure full employment and the general well-being of the country in a way which it is not easy to do in the actual terms of the Bill itself owing to the negative character of the powers'.

Gilbert was very sceptical about this twist, pointing out that the problem was not just one of drafting, because the powers in the Bill were negative. He thought that the 'long title and preamble are in fact merely pieces of rhetoric not supported by the substantive provisions'.[28] The only positive powers he could see the possibility of were either financial inducements to investment—but these would require a finance bill—or general powers of direction as in wartime—but this would be unworkable in peacetime.

Despite such scepticism, the Cabinet decided to press ahead with the preparation for the retitled Bill, and to mention it in the King's Speech at the opening of the new Parliament.[29] But it was accepted that the gap between the title and the contents of the Bill would have to be narrowed by including it in some more positive powers. This led to a wide debate about possible powers and policies to combat unemployment.

A committee under the Minister of Health essayed the possibility of two broad kinds of powers: powers to buy, sell, and manufacture by departments, and powers to encourage private investment financially.[30] Gilbert attacked both sets of proposals. He thought the first kind of power 'very wide and likely to cause disturbance or indeed alarm over a wide range of production'. He also stressed the seeming lack of any clear idea as to the circumstances and purposes of such powers.

Against the second kind of powers Gilbert enumerated several further objections. First, the role of financial powers in a depression was unclear. Second, any such powers would have to be widely drawn, 'and would be likely to cause confusion in the public mind, especially as the Government's policy as far ahead as one can see must be to educate public opinion against the risks of inflation rather than of a depression'.

Overall, he re-asserted, the only good Bill would be an economic powers Bill which dropped the title and rhetoric of full employment.[31] But this line of argument was resisted by Douglas Jay, who stressed the political undesirability of having a purely negative bill. He stressed that the Labour Party election manifesto of 1950 had committed Labour to wide powers to manufacture and purchase without causing widespread alarm. He also resisted the idea that what would be needed in a slump could not be foretold—this would run against the whole trend of policy as illustrated by the ECOSOC resolution.[32]

Gaitskell's response to this conflicting advice was to grasp the logic of Gilbert's position and recast the Bill as entirely concerned with full employment;

> The political importance of this Bill is so great that we must think out the implications very carefully before we decide what to put in it. The first step is to agree on the general case that we intend to make in presenting the Bill. The content of the Bill should then follow this. We should not put in the Bill powers which we do not regard as essential to its purpose. The means, I suggest, that the starting point should be not the Supplies and Services Act [i.e. emergency powers] but a Full Employment Policy.[33]

This evolution of the Bill towards a sole concern with full employment was conditioned by a number of factors. Positively, the ECOSOC resolution was pushing Britain towards offering a summary of its full-employment policies. Negatively, the beginnings of the Korean war completely changed the context of the discussion over powers, as the immediate need was no longer to discard wartime powers in favour of those needed for peace, but to extend the wartime powers to fight a new war.[34]

In addition, Gaitskell saw the Bill as having another important function—outlining the government's policy on inflation. So when he recommended his 'new White Paper' to his ministerial colleagues in November 1950, he put the case in a number of different ways. On the positive side, he stressed that 'We have had five years' experience in full employment in a planned economy, and some re-statement of our views in the light of this experience, which is quite new to the UK—and to other countries—would be well worth doing. I would expect such a White Paper to tell the story of how we have preserved full employment in the last five years and the lessons we have learnt in doing so.'

He also saw the usefulness of going over the various schemes for stopping deflation outlined in the 1944 White Paper, and considering any additional powers that might now be needed. However, he believed that 'the main contribution, as compared with the 1944 White Paper, is the technique of preventing inflation. Here there should be a full explanation both of the necessary monetary and fiscal policies and also the need for physical. Finally, the White Paper can deal with the probability of inflation resulting from increases in costs and incomes with the measures required to deal with it.' All this, Gaitskell felt, was of considerable value in educating public opinion. Reflecting on these considerations, he suggested that the Bill be retitled the Full Employment and Control Bill.[35]

This is not the place to examine government policy on inflation. Jones has examined in detail the position up to 1948.[36] As he points out, while economists had, from the beginning of discussion of full employment, been aware of the potential inflationary dangers, no substantial policy changes followed from this. In 1948 the government negotiated a successful wage agreement with the TUC, which lasted into 1950. But from early that year the policy came under pressure, though wages rose rapidly only at its end. The breakdown of the policy was accompanied by considerable government effort to negotiate a new arrangement. At Hall's suggestion, discussion of wages and incomes policy was put in the context of discussions of full employment. Gaitskell's policy was for some kind of national advisory body on wages, though without any compulsory powers.[37]

While the inflationary aspect was politically most prominent in late 1950 and early 1951, the preparation of the new White Paper involved a great deal of work on anti-slump measures in a number of government departments. Discussion of powers relevant to such measures had begun back in the summer of 1950, when the Ministry of Health circulated a memorandum that made the twofold distinction in powers noted above.

Gilbert was unenthusiastic about this effort. He thought the emphasis was wrong: 'I would in fact rather see a comprehensive study of how to live comfortably in a boom than how to do the same in a slump.' He went on to outline his piecemeal approach to anti-slump measures.

> The sort of powers we already have are specific. The National Insurance Act, for example, includes power to vary contributions. The Borrowing Act enables us to assist reconstruction or development of an industry. The Distribution of Industry Act enables us to assist particular areas of depression. I have always thought that further measures would have to be of the same specific character, for example, a Bill to stimulate capital investment in the socialized industries, or to give special help to local authorities. If this is a right judgement of the situation, it would in a way be unfortunate to encourage the idea by a White Paper that a single comprehensive measure of an appropriate character could be produced.[38]

Such views were powerfully represented in the committee that considered Gaitskell's new proposals, chaired by Eady of the Treasury.[39] In his sketch of the positive powers that might be appropriate for use in a slump, Gaitskell has stressed the role of fiscal and monetary measures in a slump. But he suggested four powers that might be useful:

(1) powers to stimulate public investment;
(2) powers to stimulate private investment;
(3) powers of manufacture;
(4) powers of purchase.

On (1) Gaitskell thought new powers were needed *vis-à-vis* local authorities and nationalized industries; in the latter case, 'the problem is to find some way of overcoming their reluctance because they fear that on a particular project they would lose money'.

He accepted that the issues raised by stimulation of private investment were much more difficult, but thought this an area where powers ought to be taken. However, in the case of powers of manufacture and purchase he was much more sceptical, on both economic and political grounds. He thought state powers to manufacture might defer investment by the private sector, and state purchases would raise serious difficulties of choosing commodities, deciding how they were to be sold, and choosing the organization to do this work. Overall, he stressed, 'We do not want to provoke a degree of alarm and nervousness in the private sector or of political trouble which would be quite out of proportion to any advantage there might be for maintaining full employment.'[40]

In the debates around Gaitskell's proposals, the Treasury focused on the problems of the positive powers suggested. On the one hand, Eady attacked the idea of any mention of budgetary steps: he stressed that such measures could not be constitutionally pursued in such a Bill—they would have to be the subject of a finance Bill. He was also strongly opposed to any clear specification of steps to financially stimulate investment in a depression. This was partly on the essentially political ground that to take wide powers in this area would be confusing to the public, especially as the government's task is to 'educate public opinion, and to take action, against the risks of inflation, rather than depression'.

In addition, he didn't want to encourage the view that industry only needs to ask for help and it will get it. Finally, he thought that, if a new agency for this purpose were established, it would be likely to be used inappropriately, for example for depressions in particular industries.[41]

The Economic Section was also very much concerned with the problem of inflation. Its initial brief in work in the area of the proposed legislation had been to write an assessment of the previous years' experience of employment policy, including, although not exclusively, the implications of this for wages and prices. But a draft of a report on this work was thought by Gaitskell to emphasize the wages problem exclusively. He suggested that the right balance should be 25 per cent on wages and income policy, 25 per cent on other anti-inflationary measures, 25 per cent on possible stimuli to activity in a slump, and 25 per cent in general.

Like the Treasury, the Economic Section seems to have entertained considerable doubt about the usefulness of a new Employment White Paper. Butt of the Economic Section opined in October 1950 that

> A White Paper dealing with how to maintain employment in a slump seems likely to be dull for two reasons. First, nobody believes that a slump is on the way. On the contrary, our danger in the next 3 years is surely inflation. Secondly, the Coalition White Paper of 1944 was an admirable document and there is really very little to add to it about the prevention of

unemployment. I think we could only concoct a few devices of a somewhat peripheral and academic character. It would be quite fun to do so, but the result would 'butter no parsnips'.[42]

Despite these doubts, committee discussion of the proposed Bill continued through the winter of 1950/1. In December a draft Bill was produced, by now entitled simply a Full Employment Bill. This first draft did not include the positive power to stimulate investment opposed by the Treasury—but these were inserted in the second draft at ministerial insistence.[43] Powers to subsidize local authorities and socialized industries to enable them to prepare and revise plans so as to create a reserve of works for use in a future recession were also included in the second draft, following a Central Economic Planning Staff report on reserve of works.[44]

Another sub-committee had reviewed the issue of state manufacture and purchase. As already noted, Gaitskell had been sceptical of the economic and political case for such powers. The official committee on the Bill opposed any suggestion of powers to manufacture because it would either involve building new plant, which would be too slow to be effective; or would mean taking over existing plant, which would frighten the private sector. On purchasing, the committee followed a Board of Trade argument which suggested that powers to purchase be confined to capital goods, with consumer goods being left to general fiscal policy.[45]

All in all, the Full Employment Bill went through three drafts: December 1950, and January and February 1951. Through those re-draftings the powers envisaged had been strengthened. Nevertheless, the Treasury view on the problems of such powers is very evident to the end. In particular, their unhappiness at any subsidizing of private investment is apparent in the conditions accompanying such loans, as outlined in the third draft. These conditions were threefold: that

(a) the giving of financial assistance for the purposes of any undertaking carried on or proposed to be carried on in the UK is likely to contribute to the increase or maintenance of employment in the UK;

(b) such assistance would not be available on reasonable terms from any other source; and

(c) there are reasonable prospects of the undertaking ultimately being able to be carried on successfully without further assistance under this section.[46]

This hardly amounted to a programme likely to avert a major decline in private investment. And overall, the new powers envisaged by the Bill fell well short of a comprehensive programme to avert a depression despite the view in the Treasury that it involved many 'novel and controversial features'.[47] But by early 1951 such a depression appeared a remote possibility: 'any danger to full employment through a slackening of the demand for goods and services has been relegated to a relatively distant future. If employment is in danger, it is through a scarcity of supplies. A Bill which was designed to stimulate demand would not be regarded by Parliament or the country as relevant to our current problems.' The reason for this change of circumstance was of course the Korean war, and this too made the question of economic powers for peacetime—the original stimulus to the whole process—beside the point, at least in the short run. Hence, in February 1951 the Cabinet decided to drop the idea of introducing any Bill for the foreseeable future, and none indeed was introduced before the fall of the Labour government.[48]

Conclusions

In 1950, the fear of a domestically generated depression had largely disappeared, at least in the foreseeable future. Fears of unemployment arose mainly from doubts on Britain's ability to export sufficiently to finance a full-employment level of imports, and the danger here was seen as likely to be gravely exacerbated by any US recession. Thus the emphasis was on encouraging the United States to take measures to lessen the likelihood or impact of any such recession—especially in years like 1949, when they did have a mild recession. This view of the source of dangers to Britain's employment level led, as noted above, to the 3 per cent target 'pour encourager les autres'.

Such targeting ran against the Treasury's dislike of seeming to accept that the cure for almost any unemployment was more aggregate demand. The Treasury continued their 1920 and 1930s role of emphasizing the structural and frictional components to such unemployment. In addition, the commitment to 3 per cent may seem to have raised the problem of inflation. The Treasury certainly viewed inflation very seriously, and accepted the view that there was a trade-off between unemployment and inflation;

but they took the view that a relatively small increase from the prevailing levels of around 2 per cent would be sufficient to cope with this danger. In slightly anachronistic terms, they seem to have envisaged a Phillips curve with a very steep slope. Hence there is a surprising lack of resistance to a low target level of unemployment, even if officials would have preferred 4 per cent to the eventual 3 per cent.

The remoteness of the threat of unemployment by the late 1940s meant that, as far as involvement with private investment was concerned, either administrative or financial, the Treasury was reluctant to concede much ground. On the one hand, it didn't want to get involved in any detailed way itself; on the other hand, it did not want any separate agency with wide powers which might frighten the private sector.[49] Reluctantly, it retreated under ministerial pressure to consider financial subsidy, but hedged around with strong qualifications. Equally, it resisted the idea of any automatic measures against slump, maintaining the view that Treasury discretion was vital to assess all the circumstances before action was taken (see Chapter 5 above).

At the ministerial level, the enthusiasm to proclaim success on full employment was tempered by fears of inflation. And while it might make sense to economic advisers to try to link full employment to inflation in negotiations with the unions, the Labour government made little if any suggestion that inflation threatened full employment. Such a claim not only would have been politically difficult to sustain, but also would have found no support in current economic arguments which clearly saw a link from full employment to inflation, but did not argue that the harms from inflation included a reduction of unemployment.

By 1951, therefore, it was generally accepted that unemployment arising from domestic sources was unlikely, and that in any event could be fairly easily dealt with.[50] The fears of 1945–7, of a boom followed by a slump, had disappeared. The one remaining danger was seen as an American recession, and this continued to worry governments well into the 1950s (see Chapter 8 below).

Notes

[1] Announced in the House of Commons March 1951, Debates 1950/51, Vol. 485, Col. 319–20.

[2] E.g. by James Meade: PRO T230/69, *A Comparison of Beveridge's Full Employment in a Free Society and the White Paper on Employment Policy*.

[3] R. N. Gardner, *Sterling–Dollar Diplomacy* (Oxford, 1956), 37.

[4] J. M. Keynes, *Activities 1940–44: Shaping the Post War World: The Clearing Union* (Collected Writings, 25; London, 1980), para. IV, 18.

[5] Gardener, *Sterling–Dollar Diplomacy*, 96–7.

[6] J. Meade, Diaries (November/December 1944).

[7] J. Garratty, *Unemployment in History: Economic Thought and Public Policy* (New York, 1978).

[8] Gardner, *Sterling–Dollar Diplomacy*, 146–7.

[9] Ibid., 271.

[10] Ibid., 379.

[11] PRO T230/230, *Economic and Social Council—11th Session: Report by M. Fleming* (August 1950), para. 1.

[12] PRO T230/230, B. Gilbert to M. Fleming (4 October 1948).

[13] Summary of this report in PRO Prem 8/1189; see also PRO T229/432 and T229/433.

[14] PRO CAB 134/22, *Report by Interdepartmental Working Party on United Nations Report on National and International Measures to Achieve Full Employment* (31 January 1950); earlier drafts in PRO T229/432.

[15] PRO CAB 134/22, para. 7(b)(ii).

[16] The experts had proposed quantitative targets for foreign investment, to be made good by public investment if necessary. They had also suggested that any country reducing its imports by failing to maintain domestic demand should deposit currency to the value of the shortfall with the IMF. Both proposals were regarded as wholly impracticable in the Treasury; see e.g. PRO T229/432, Draft Report, *The Full Employment Standard* (26 October 1950).

[17] PRO CAB 134/225. *Report on the Fifth Session of the Economic and Employment Commission* (New York, January 1950) . . . by M. Fleming (10 February 1950). See also CAB 134/225, *United Nations Report on National and International Measures to Achieve Full Employment: Note by the Economic Section* (undated).

[18] Ibid., paras. 9(a) and (b).

[19] Fleming, *Report on the Fifth Session*, para. 11; see also PRO 229/432, R. Hall to Minister of State for Economic Affairs, *Report of Measures to Achieve Full Employment* (27 June 1950).

[20] PRO T230/230. *Summary of Ministers Comments on Working Party's Report* (27 January 1950).

[21] PRO T230/230, *Economic and Social Council—11th Session*, para. 13b.

[22] PRO CAB 134/263, *Economic Steering Committee, Minutes* (7 December 1950).

[23] PRO CAB 134/229, *Memorandum by the Chancellor of the Exchequer on the Full Employment Standard, Annexe* (1 January 1951).

[25] The only previous use of a set percentage target in this period was the use of 5 per cent (male) unemployment as triggering an easing of cuts in investment programmes: PRO CAB 134/438, *Investment Programmes Committee, Minutes* (13 April 1948).

[25] E.g. PRO T230/243, *Economic Steering Committee, Minutes* (18 December 1952).

[26] PRO T229/266, *Official Committee on Economic Policy and Full Employment Bill, Minutes* (4 July 1950).

[27] PRO CAB 128/17, Cabinet Conclusions (30 March 1950).

[28] PRO T228/241, Memorandum by B. Gilbert (26 July 1950).

[29] PRO CAB 128/18, Cabinet Conclusions (19 October 1950).

[30] PRO T228/241, *Memorandum by Minister of Health on Economic Planning and Full Employment Bill* (17 October 1950).

[31] PRO T228/241, *Economic Planning and Full Employment Bill* (26 October 1950).

[32] PRO T228/241, Jay to Chancellor of Exchequer (27 October 1950).

[33] PRO CAB 130/60, GEN 343/2, *Economic Planning and Full Employment Bill: Memorandum by the Chancellor of the Exchequer* (9 November 1950).

[34] PRO T229/267, *Report by the Official Committee on the Economic Planning and Full Employment Bill* (31 January 1951).

[35] PRO CAB 134/227, *White Paper on Full Employment: Note by the Chancellor of the Exchequer* (8 November 1950); CAB 134/224, *Economic Policy Committee, Minutes* (10 November 1950).

[36] R. Jones, *Wages and Employment Policy, 1936–86* (London, 1987).

[37] A. Cairncross, *Years of Recovery: British Economic Policy 1945–51* (London, 1985), 404–8.

[38] PRO T228/241, Gilbert to Bridges (18 October 1950).

[39] PRO T228/241, B. Trend to W. Eady (18 October 1950).

[40] PRO CAB 134/227, *White Paper on Full Employment*, paras. 7, 11.

[41] Memorandum by Eady (no date, but probably October 1950).

[42] PRO T229/323, Butt to Hall (27 October 1950).

[43] PRO T229/267, *Official Committee on Economic Planning and Full Employment Bill, Minutes* (10 January 1951).

[44] PRO T228/244, *Report by Working Party on Reserve of Works* (26 January 1951).

[45] PRO T228/243, *Draft Report of Sub-Committee on Power to Manufacture, Purchase and Sell Goods* (31 January 1951).

[46] PRO CAB 130/60, *Full Employment Bill*, para. 2(1). (Other drafts are in the same file.)

[47] PRO T228/242, *Economic Powers and Full Employment: Proposed Legislation*, Memorandum (8 February 1951).

[48] PRO T228/242, *Full Employment Bill: Memorandum by Lord President of the Council* (27 March 1951).

[49] PRO T228/242, A. Grant to B. Trend (5 February 1951).

[50] It is interesting to note that the discussion of the employment standard was seen as having virtually no *domestic* economic significance. One commentator had suggested that such a target might maintain confidence among domestic investors, but this point seems to have aroused little support. PRO T229/432, J. V. Licence to R. Hall, *Report on National and International Measures for Maintaining Full Employment* (23 January 1950), para. 8.

8

Employment Policy under the Conservatives: Butskellism, 1951–1955

One of the major reasons for the stunning defeat for the Conservatives in the election of 1945 was their association with the mass unemployment of the 1930s. During their period in opposition, 1945–51, they attempted to erase this association by making a wholehearted commitment to full employment as a policy objective. Thus, in the general election manifesto of 1950 the Conservatives emphasized that 'We regard full employment as the first aim of a Conservative Government.' The only caveat to this was the problem of a cessation of US aid. In 1951, 69 per cent of Conservative election addresses made promises on full employment.[1]

The years in office of the Conservatives in the early 1950s were marked by a continuation of the unprecendentedly low levels of unemployment inherited from the Labour government. This fact, the Conservative commitment to full employment, and the lack of sharp change in economic policy with the replacement of Labour's Gaitskell by the Conservatives' Butler as chancellor of the Exchequer, led to the term 'Butskellism'. This was used to describe the alleged consensus on economic policy between the two chancellors, and their commitment to 'modernization' against the 'wilder' elements in their own parties.[2]

There is no reason to doubt the strength of the political calculation by the Conservatives that a commitment to full employment was electorally inescapable. By the 1950s this calculation had become embedded in British politics, and was indeed to remain so until the early 1980s. In the early days of Butler's chancellorship the view was expressed very forcefully by Plowden, the chief of the Economic Planning Staff and a crucial government economic adviser: 'My own belief is that the British people have an almost pathological fear of unemployment and therefore, unless it is demonstrated to them that it is insoluble or not as bad as they fear, any political party must continue to maintain "full employment" as its aim.'[3]

Plowden went on to argue that such a policy commitment was definitely not compatible with a free market economy, and that the government must therefore continue to use a wide variety of instruments to manage the economy. In this way, he demonstrated the incompatibility of full employment and some of the wilder claims for *laissez-faire* policies to be heard in the Conservative Party. But for the majority of that party it was apparent that, for the foreseeable future at least, a full-employment commitment did not involve any very painful policy choices. In other words, such a commitment could be easily given when its pursuit did not involve any serious sacrifice of other policy goals. As will be argued below, when such a sacrifice was suggested as a price for full employment policy, the willingness to bear it is far from apparent.

Linked to this point, the fact of full employment under the Conservatives cannot of itself tell us much about their commitment to that end. They inherited from Labour a position of excess demand, and while this came to an end in 1952/3 they were never assailed by strong deflationary forces. Hence they were never called upon in practice (as opposed to policy discussion) to test this commitment.

The continuation of full employment under the Conservatives was rendered largely unproblematic by the 'friendliness' of external circumstance. World output and trade growth continued unabated during this period, with output growing at an average rate of over 4 per cent per annum in the 1950s, and trade even faster.[4] The recession in the United States of 1953/4, while arousing much fear in Britain (see below), was in fact largely offset in its effects on British trade by the very rapid growth of European countries at this time, most notably West Germany.[5]

It is in this context that we must look at the realities of Butskellism: first, by looking at the actual conduct of policy, the arguments surrounding its formulation, and the implications of these for employment policy; second, by looking at the voluminous and wide-ranging debates about what to do in the event of a recession. While these discussions, in policy terms, eventually came to naught, they are highly instructive on the state of debate about policy during the Butler years.

Policy 1951–5

The central theme of Conservative economic policy announcements around the 1951 elections was to free the economy of controls. This process had already begun under Labour, and in fact was reversed by the Conservatives in late 1951 and early 1952 in the context of the Korean war emergency,[6] but in the longer run the trend was re-asserted. The public records suggest that talk of 'freeing the economy' was not just a rhetorical weapon of the Conservatives, but a strongly held belief about the proper functioning of the economy. It was an issue that was highly charged in the discussions of policies against a slump (see below).

In the realm of policy instruments rather than objectives, the Conservatives were strongly committed to the restoration of monetary policy (Bank rate), in opposition to the 'cheap money' policies of the previous two decades. The basis for this commitment was several-fold. In part, it derived from a concern with inflation and with the belief that monetary policy had a part to play in its control.[7] Second, monetary policy was seen, especially by economists, as one among several weapons for managing the economy which had been dogmatically rejected by the previous government.[8] Third, monetary policy was regarded as part of the re-creation of a policy framework in which the balance of payments would act as a barometer of external confidence in the conduct of British domestic policy, and where the Bank rate would be the primary instrument of response to such external judgements.[9] While policy was never so dominated by a single measure or instrument as this type of argument suggests, the question of financial confidence was indeed resurrected under the Conservatives to a key role in policy-making. Again, the issue came to be of particular importance when measures against a threatened slump were under discussion.

But confidence also bore on the discussions of fiscal policy, which remained the major instrument of economic management; Bank rate was used under the Conservatives, but for most of the time, at least, in subordination to the broad thrust of policy determined in the Budget.[10] In the making of those Budgets, confidence did not by any means lose the role it played in Budget-making before the war, and as discussed in Chapter 3 above. But this is best considered in the context of a general outline of budgetary policy.

Central to Conservative concern with budgetary policy was a desire to reduce both taxation and expenditure. They succeeded in the former but not in the latter. Such an outcome was compatible with no sharp increase in government borrowing, because the period began with very large budget surpluses, which gradually fell in size as private savings rose, increasing by over 200 per cent over 1948–53.[11] Cuts in government expenditure were continually emphasized as central to government policy, but the most that was attained were cuts in previously announced rates of growth. Nevertheless, fiscal policy was as much about the search for public expenditure economies as it was about the fiscal balance.[12]

Table 2. Public Expenditure and the Budget, 1939–55 (£m)

	Public expenditure (a)	Above-the-line surplus or deficit (b)	Savings of public sector* (c)
1938/9	940	−13	—
1939/40	1325	−276	−477
1940/1	3884	−2476	−2093
1941/2	4776	−2702	−2768
1942/3	5637	−2818	−2844
1943/4	5799	−2761	−2980
1944/5	6062	−2824	−2871
1945/6	5491	−2207	−2358
1946/7	3927	−586	−348
1947/8	3209	636	249
1948/9	3176	831	662
1949/50	3375	549	757
1950/1	3258	720	869
1951/2	4053	380	830
1952/3	4351	88	606
1953/4	4275	94	509
1954/5	4305	333	534
1955/6	4496	397	780

*Calendar years, from 1939.

Sources: Cols. (a) and (b): 1938/9–1944/5, *Annual Abstract of Statistics* (1935–46); 1945/6–1950/51, *Economic Trends* (December 1961); 1951/2–1954/5, *Annual Abstract of Statistics* (1956); Col. (c): 1939–1945, *Annual Abstract of Statistics* (1935–46); 1946–1955, *Economic Trends*, Annual Supplement (1981).

Inheriting a serious balance of payments crisis, the Conservative government had already tightened monetary policy and reintroduced certain physical controls before the Budget of 1952. Somewhat surprisingly, given the air of crisis within which it was introduced, the Budget was broadly neutral in intent in fiscal terms, though bearing more heavily on investment in an attempt to free resources for the balance of payments. The aim was, broadly, to reassure foreign opinion while not deflating demand at home to any substantial extent. The further increase in the Bank rate was aimed primarily at the former, the neutral Budget at the latter.

Such a picture is apparent from the standard sources on policies in the period—Dow, and Worswick and Ady. What the public records suggest is the extent to which there was a lack of clear strategy on overall economic policy, and a dominating desire to restore external confidence in British policy, the restoration of the gold reserves being seen as the *sine qua non* of policy.[13] To achieve this latter objective, there was pressure for some fiscal as well as monetary deflation—but this was heavily qualified. 'A moderate degree of unemployment would be welcome provided that it were moderate and provided that it were only transitional.'[14] In the event, budgetary and fiscal policy were not co-ordinated, but attempted to achieve two separate objectives.

The 1953 Budget is much more significant in relation to employment policy. Some have emphasized that this was the first Budget that aimed at expansion to counteract the growth of unemployment.[15] Certainly, for the first time since the war the economic management problem was clearly not one of excess demand, although this was rather uneven between sectors. Certainly, also, the economy was stimulated, and the objective was to prevent unemployment from rising. Within rather narrow parameters, this is an important change; but those parameters *were* narrow ones, and, most importantly, they involved no movement into budget deficits to stimulate the economy. Arguments over the Budget suggest that, if such a deficit had been required, it was unlikely to have been forthcoming.

In February 1953 Hall argued that, on the basis of the normal calculation of the above-the-line budgetary position (see below for discussion of the meaning of this), there was a case in the forthcoming Budget for a reduction in the 1952/3 surplus or even for some deficit, in order to stimulate the economy. But, he went on, 'the balance of payments position is still precarious, and it is much easier to give money away than to get it back again if circumstances change. It seems likely also that the psychological effect of moving to an above-the-line deficit this year would be very damaging. If this view is accepted, it sets a limit to the margin available of say £150m.'[16]

This view was accepted by other members of the Budget Committee, and indeed some felt that not enough regard had been paid to the external position. Bridges argued that, if the sole concern were employment policy, the fiscal stance advocated by Hall was correct, but 'it was clear from the monetary point of view and from the balance of payments point of view, [that] we must endeavour to keep a certain amount of slack in the system'.[17]

In part, the issue was one of a fairly narrow dispute over the precise degree of fine-tuning of the level of demand. But Hall's memo raised more fundamental issues about the conduct of budgetary policy. Hall himself had raised the issue of confidence in his paper, but in discussion he also stressed the Keynesian line that the budgetary position should be judged by reference to the overall level of demand required. This was unacceptable to many in the Treasury, who emphasized the traditional Treasury opposition to budget deficits. The draft budget speech which evolved from the Budget Committee deliberations included the sentence, 'I consider that it would be out of accord with the spirit of the Commonwealth Conference, bad for confidence, and indeed, unsound in itself to budget for any deficit above-the-line.'[18]

Hall, as an advocate of the Keynesian use of the Budget, even if that involved deficits, found himself in a difficult position. In 1953, and again in 1954, he supported budgeting for a surplus above the line, but he wanted to make clear that this was based on his assessment of the level of demand rather than on attachment to surpluses *per se*. Thus, prior to the 1954 Budget he emphasized to the Budget Committee 'that although he would advise against concessions this year it would not be on the grounds that they would produce a deficit above the line, but because he thought that the present budgetary position was about right from the economic stand point'.[19]

In stating this principle, Hall was in a minority not only among the civil servants of the Budget Committee, but also in arguing against the Chancellor's approach. At the same meeting as Hall made the above statement, Butler stressed the need to reduce public expenditure, and made clear that he was 'determined not to have a deficit on the budget'.[20] A fortnight later Butler made the same point in relation to the possibility of a deficit in the 1955/6 Budget.[21]

The basis of this opposition to prospective deficits was then twofold. First, and predominant, was the fear of the effects on confidence, which as noted above was given a great weight by this government in all its policy deliberations. As in the 1930s, such an emphasis could of course be purely self-serving—a blind to cover actions taken for other reasons. To some extent, also, 'confidence' is a circular argument; for, once governments argue that financial deficits deserve a loss of confidence, the markets are likely to respond precisely in that manner to any such deficits.

This emphasis on confidence could be made intellectually respectable by posing the issue as one of an unfortunately unenlightened world, which however had to be lived with. Thus, in a classic minute which could have been written at any time since the 1920s, Rowan wrote:

> As regards the question of a surplus above-the-line, this is not of course, I agree, a major matter. But on a secondary plane it has a good deal of psychological importance as it inspires confidence. It may be that all this is old-fashioned, but we have to deal with the impression which we make on people as they are, and, unfortunately, not on people who are really versed in some of the deeper methods of Budget planning.[22]

Second, opposition to deficits was based on the desire of the Treasury to control government expenditure. As noted above, the government had committed itself to public expenditure cuts, and although these were not forthcoming there was constant searching for ways of at least reigning back any expansion. Budget deficits, it was argued, 'would be taken by the spending Departments as a sign that the Treasury was prepared to relax'.[23]

Hall responded to these lines of argument by pointing to the 1944 Employment White Paper and Cripps' budget speech of 1948, with their endorsement of budgeting for a balance only over a period of years. He argued strongly that it would be retrograde to give the impression that a surplus was being budgeted for any reason other than its desirability for the management of the level of demand.[24]

These disputes over the use of the Budget never came to a 'crunch' because both sides were agreed that the desirable policy in the current circumstances was a budget surplus. It also has to be said that Butler, while usually hostile to deficits, was also strongly influenced by Hall's approach, so an unequivocal statement of principle was never publicly forthcoming from the Treasury.

The issue was also affected by the problem of budgetary calculation. As is apparent from the above discussion, budgetary analysis at this time was based on the traditional classification of the budget accounts. This divided them into above-the-line and below-the-line accounts on the basis of whether or not parliamentary approval had been given for borrowing, and not on the basis of notions of current and capital expenditures, or of items to be financed by current revenue or borrowing.

Of course, since the early years of the war, accounts had been prepared which had integrated the Budget with the national accounts in order to measure the overall pressure of demand. But in the 1950s these calculations (published in the National Income White Paper) were treated as separate and subsidiary to the traditional form of accounting.

In 1948 an Alternative Classification had been included in the budget statement, which moved away from the traditional method to some degree, but still focused entirely on Exchequer transactions (ignoring, for example, insurance funds) and which made only a limited separation of capital and current accounts. A full 'economic classification', based on the same principles as the National Income White Paper, was devised and used in the budget discussions of 1952/3, but was not published. In this and succeeding years there was a lot of discussion about the desirability of publishing this series, but pressure to do so was successfully resisted.

This Treasury resistance was argued on two main grounds: first, that the orthodox cash Budget was an instrument of control that would be weakened if another form of budget calculation were published alongside, and particularly one that might affect confidence by its different figure for the surplus or deficit; second, that the complexities of calculation of the economic classification would lead to much greater scope for questioning of the Chancellor's budget judgement.[25]

This defence of the traditional budget accounts obviously went along with a concern for the traditional functions of the Budget, namely, to control expenditure and to maintain confidence in the conduct of policy. Thus there was an uneasy compromise in this period between an undoubtedly new use of the Budget—to regulate demand—and an attachment to traditional functions and procedures. This did not lead to any great problems at the time, because the budgetary position was underpinned by the buoyancy of demand from private sources.

Leaving aside this internal quarrel, the actual conduct of budget policy was one of fine-tuning demand, but within the constraints of a successful desire to cut taxes, and an unsuccessful desire to cut public expenditure. The first of these was important in the reluctance to use tax increases as a short-run stabilizing device, and one ingredient of the highly expansionary Budget of spring 1955.[26] Another ingredient in this was no doubt an exaggerated belief in the capacity of monetary policy to offset fiscal policy. But above all, the early 1955 Budget was significant for employment policy because for the first time it saw the manipulation of a demand increase to help the incumbent government win re-election, against the economic advice which called for more restrictive policies. Once the election was won, the restrictions came.[27]

Such manipulations of demand focused on the aggregate level of spending rather than on the components. Of course, it had to have regard for the balance of payments position, and this laid the basis for what later came to be called stop–go, with its concern alternating between a deteriorating balance of payments and a rising unemployment rate. But even within domestic demand, there was a strongly expressed desire to increase investment while holding back consumption once the general pressure on investment resources slackened after 1952.[28] In practice, this concern led to very little action. After a fight against traditional Inland Revenue conservatism on the use of tax devices for macroeconomic purposes, Hall carried a proposal for investment allowances into the 1954 Budget. These replaced the initial allowances, which had been restored in 1953, with what amounted to permanent tax exemptions.[24]

However, this regulation of the level of investment was largely *ad hoc* and was not tied to any clear appreciation of the determinants of investment levels. Attempts to use investment as a regulator of demand had acquired a bad name in Britain, but in part this was a function of the unreflected nature of the policy. Some interest was taken in the much more systematic Swedish schemes of reserve funds, but such a complex use of the tax system to regulate the national economy was unlikely to have an easy time against Inland Revenue conceptions of the appropriate functions of taxation, though it remains unclear why it did not at least appear on the British policy agenda.[30]

To summarize on budget policy and employment in this period is to emphasize the speculative against the actual. In the event, maintenance of employment never posed any serious policy difficulties at this time. The economy could be readily fine-tuned in a world economy that provided buoyant conditions for policy to operate upon. The budget debates within the government must raise some doubts about what might have happened if a serious slump *had* occurred. But this question raises in even sharper form by the discussions that went on precisely to consider such an eventuality.

Measures against a Possible Recession

Like its Labour predecessor, the Conservative government after 1952 saw a significant danger of a recession originating abroad, especially in the United States. This fear was in general terms occasioned by the memories of the slump of the 1930s and an appreciation of the weight of the US economy in world trade, and hence the effects of any downturn there on the British economy. In addition, there was a fear that the diffusion of powers in the US governmental system would militate against any rapid response to a domestic recession, whatever the formal intentions of the federal executive.

From 1952, these general fears were fused with a more specific worry that the end of the Korean war boom would lead to a recession. 'Since the war, activity in the US has been sustained first by pent-up demand carried over from the War and then by the Korean boom and the subsequent expansion of defence expenditure . . . the danger now is that a recession will occur as defence expenditure flattens out and as investment in defence expenditure declines.'[31]

In response to such worries, a Cabinet committee was set up

chaired by the Chancellor of the Duchy of Lancaster (Lord Swinton), 'to formulate and submit to the Cabinet practical plans whereby effective remedial measures could at once be applied to counter any tendency towards widespread unemployment'.[32] The committee took as its measure of 'widespread unemployment' the 3 per cent figure agreed by the previous government, but not formally endorsed by the Conservatives. By this standard, they saw no likelihood of serious unemployment in the near future, but expressed fears about the possibility of a US recession.[33] However, the committee stood by for future references, and in 1953, as the fears about the US economy were to a degree realized, the volume of work on the possible effects of such a recession multiplied. Much of this work took place under the aegis of an Economic Steering Committee Working Group on Employment Policy established in November 1952. This in turn divided most of the work into domestic and international aspects.[34]

From the beginning, work on domestic responses to a recession was dominated by the question of a reserve of works in the public sector. Initially, the enquiry was into how for such a reserve already existed as a natural by-product of the pent-up demand for investment in the public sector. Once the investigation began, exactly the same issues came to the fore as in the previous discussion of this topic. Local authorities and other public sector agencies did not have a large reserve of projects, because their personnel were fully occupied with current work, and because of the costs incurred in such pre-planning and the pre-purchase of land usually entailed by such projects. Hence, as had happened twice under Labour, a committee was set up to look at the desirability and difficulties of a reserve of works scheme.[35]

This focus on reserve of works to combat a recession was problematic from the start. From the beginning, Gilbert from the Treasury had opposed any central government money for such programmes.[36] This opposition remained constant throughout all the subsequent discussions. But in addition, such schemes ran up against the desire of the government to cut back on non-remunerative public investment, and to encourage investment by nationalized corporations and private industry.

This combination of objections meant that the Report of the Working Group on Employment Policy very much played down such schemes. It emphasized that they would be only a small part

a warning note was sounded: 'We have, thus, much more experience of using some of the instruments of employment policy, though it must be remembered that we have not yet been faced with the problem of a serious slump.'[42]

The issue of a prospective slump generated much more of a clash of policy approaches on the international side than on the domestic. A committee to look at this was set up chaired by Plowden. The papers presented to this committee showed diametrically opposed views on crucial issues, from the Economic Section, the Board of Trade, and the Paymaster General on the one hand to the Overseas Finance Section of the Treasury on the other. The former group advocated that in the event of a recession a small devaluation to a lower exchange rate might be desirable, but the main response would be the rebuilding of import controls and the establishment of protected trade blocs with both the Commonwealth and European countries to offset the loss of US markets. Against this, the Treasury advocated a floating exchange rate and exchange convertibility as the only alternative to deflation—basically, the Robot package of 1952.[43]

The Economic Section and its allies believed that a float would not solve the balance of payments problem because of the initial adverse effects (the J-curve effect) and because of the danger of a downward spiral of the exchange rate and the balance of payments that might be unleashed. They also argued that convertibility would lead to trade discrimination against Britain, as other countries sought pounds with which to purchase freely available dollars. They argued that convertibility, while desirable as a long-run aim, could come about only in a friendly international environment and with US support—not in a recession.

Against this, the Overseas Finance Section argued that the balance of payments would be righted by a floating rate, and convertibility would maintain confidence in the pound.[44] Their main concern was the balance of payments and the strength of sterling, while their opponents started from the need to maintain demand and employment.[35]

On this issue the Chancellor was inclined to side with the Treasury Overseas Finance division. First, he had of course been a proponent of the floating rate and convertibility scheme in the previous year. More specifically, he disliked several features of the proferred alternatives in the event of a recession. He saw an 'insuperable difficulty' in going for a devaluation to a lower fixed rate, which would involve a complete reversal of a policy. He did not like the idea of a re-introduction of controls, which ran exactly counter to a primary objective of his policies. Finally, he was opposed to a discrimination against the United States, especially given the need for political unity in the West.[46]

Of these three issues, the second posed an immediate question:- how far should the policy of de-control be pursued pending the impact of the US recession? A working party of officials was established to study the issue of de-control and its implications for policy in a recession. In their report they emphasized that the de-control process 'may make it more difficult for the Government to maintain a high level of activity and employment in the event of a recession'.[47]

But ministerial opinion was opposed to any slowing down of de-control,[48] and nothing was done. In the same manner, exploratory talks with Canada, the United States, and the Commonwealth were held, but no policy decisions were taken that might imply trade discrimination in the event of a recession.[49]

In the event, all those discussions about a US recession seriously affecting the UK proved unnecessary. While the United States did have a recession in 1953–4, its impact on Britain was small, mainly because of the offsetting spurt of expansion in Europe.[50] Nevertheless, the debates on how to deal with the possibility of recession are highly instructive.

On the domestic front, the focus on a reserve of works as a preparation for depression, which had been a central proposal since at least the 1930s, was all but killed off in the early 1950s. Yet it was far from clear what would replace it in the event of a slump. The efficacy of general budgetary policy in fine-tuning the economy cannot be conclusive evidence of its efficacy in less auspicious circumstances.

Equally, the trend of international policy, and especially policy on de-control and relations with the United States, made the likely international response to a slump very unclear. And even if a floating exchange rate were a nostrum whose time had not yet come, there was a great depth of hostility to devaluation, which would have been seen as a hateful repetition of that under Labour in 1949, and a sickening blow to the desire to restore confidence in sterling.

Conclusion

In his discussion of Conservative chancellors of the 1950s, Dow rightly emphasizes that, despite significant pressures for a markedly different policy from Labour, they 'in fact put full employment first as the main object of policy',[51] The material discussed in this chapter does not deny that conclusions about what actually occurred in the early 1950s.

What can, however, be drawn from this material is some doubt about the extent to which full employment would have been maintained if it had required the use of radically different policies from those actually pursued. Of course, politically, full employment had come to occupy a place never attained before the Second World War. The pressures on government to maintain that objective against deflationary forces would have certainly been very powerful. But looking back from the 1980s, it is reasonable to note how far the consensus on the centrality of full employment to public policy wsa dependent precisely on its ease of attainment. More specifically, the evidence of policy-makers' views at this time does not suggest an overwhelming willingness to sacrifice other policies to employment—whether it be fiscal rectitude, multilateral free trade, or pro-Americanism. At the minimum, there remains scope for doubt about how far commitment to full employment would have survived a severe test in the 1950s.

Notes

1 The Conservative Manifesto, 1950, in F. W. S. Craig, *British General Election Manifestos 1918–66* (Chichester, 1970); D. E. Butler, *The British General Election of 1951* (London, 1952), 55.

2 *The Economist*, (1954), 440–1: 'Whenever there is a tendency to excess Conservatism within the Conservative Party—such a clamour for too much imperial preference, for a wild dash to convertibility, or even for a little more unemployment to teach the workers a lesson—Mr Butskell speaks up for the cause of moderation from the Government side of the House' (p. 441). See also Chapter 2 above.

3 PRO T229/323, E. Plowden to Chancellor of the Exchequer, *The Government and Full Employment Policy* (10 April 1952).

4 A. Maddison, *Economic Growth in the West* (London, 1964), 28.

5 Ibid. Output grew at an average rate of 7.6 pe cent per annum in Germany in the 1950s.

[6] On the consumer side, clothing was de-rationed in 1949, petrol in 1950, various food items from 1948 onwards. By 1949 consumer rationing covered only approximately 12 per cent of expenditure. Import controls also fell away rapidly in the last years of the Labour government—by 1951 about 45 per cent of imports were entirely free from controls. Controls over investment operated mainly by building controls and steel allocation, were generally loosened under Labour, tightened in 1951/2, and then fell eventually into insignificance by the mid-1950s. J. C. R. Dow, *The Management of the British Economy 1945–60* (Cambridge, 1965), Ch. VI.

[7] Ibid., 67.

[8] PRO T229/323, E. Plowden to Chancellor of Exchequer, *The Government and Full Employment Policy* (10 April 1952); C. Kennedy, 'Monetary Policy', in G. D. N. Worswick and P. H. Ady (eds.), *The British Economy in the 1950s* (Oxford, 1962).

[9] Dow, *The Management of the British Economy*, 67–9.

[10] This role of the Budget was summarized by Hall, chief economic adviser to the government, right at the beginning of its period of office. Budgetary policy, he argued, 'cannot of itself deal with an inflation of the monetary system, or with a failure of production. Nor, unless it is pushed to the point of severe unemployment, can it prevent an upward movement of wages. But without the climate induced by budgetary methods, all the other objectives of policy become much more difficult to carry out.' PRO T171/422, R. Hall, *Budgetary Policy* (October 1951).

[11] Calculations of the budgetary position was controversial in this period, as is discussed further below, but the outcomes were as shown in Table on page 145. The change in the environment of budget-making is emphasized in PRO T230/343, R. Hall, *The Problem of Economic Expansion* (August, 1954).

[12] E.g. PRO T171/413, B. Gilbert to Armstrong (16 September 1952), arguing cuts in defence and local authority spending as priorities in preparing the budget.

[13] PRO T171/410, *Draft Budget Speeches 1951*.

[14] PRO T171/409, B. Trend, *Credit Policy* (15 February 1952).

[15] D. Winch, *Economics and Policy: A Historical Survey* (London, 1972), 299.

[16] PRO T171/430, *The Economic and Budgetary Problem in 1953* (9 February 1953).

[17] PRO T171/431, Budget Committee (11 February 1953).

[18] PRO T171/433, Draft Budget (4 March 1953). The Commonwealth Conference had emphasized the desirability of strict budget policy—PRO T171/432, R. Hall to E. Bridges, *Budget Target 1953* (7 January 1953).

[19] PRO T171/437, Budget meeting (5 February 1954).

[20] Ibid.

[21] Ibid. (17 February 1954).

[22] PRO T171/446, L. Rowan to E. Bridges, *Above-the-Line Balance* (17 March 1954).

[23] PRO T171/437, Budget meeting (1 March 1954).

[24] PRO T171/466, R. Hall to E. Bridges, *Above-the-Line Balance* (16 March 1954); T171/449, Budget meeting (15 December 1953).

[25] PRO T171/437, *The Form of Budget Accounts* (no date—but February 1954); T171/461, *Budget Meeting* (25 January 1955).

[26] C. Kennedy, 'Monetary Policy', in Worswick and Ady, *The British Economy in the 1950s*, 302.

[27] PRO T171/464, Budget meeting (9 March 1955); T171/456, Budget meetings (28 July–22 September 1955).

[28] E.g. PRO T171/437, R. Hall, *The Budgetary Problem in 1954* (27 February 1954).

[29] Ibid., Budget meeting (5 March 1954); T171/439, R. Hall, *Incentives for Investment* (11 February 1954).

[30] PRO T230/185, R. F. Bretherton, *Control of Investments by Companies and Taxation* (1 September 1950).

[31] PRO T230/271, Economic Section, *Problems and Policies in a US Recession* (9 April 1953); also T230/229, *Preparations Against a US Recession: Main Outline of a Policy* (27 April 1953).

[32] PRO CAB 134/794, *Cabinet Employment Committee*, Minutes (20 October 1952).

[33] Ibid.

[34] PRO CAB 134/890, *Economic Steering Committee Working Group on Employment Policy Meeting* (14 April 1953).

[35] Ibid., *Report on Reserve of Works* (May 1953).

[36] Ibid., Meeting (14 April 1953).

[37] PRO CAB 134/848, *Studies of Measures to Combat a Recession in the US and to Strengthen the Economy* (2 October 1953).

[38] PRO T230/271, E. Plowden to Chancellor of the Exchequer (9 October 1953).

[39] PRO CAB 134/1185, *Committee on Trade and Employment*, Meeting (29 May 1953).

[40] PRO CAB 134/851, *Memo by Minister of Works on Reserve of Works* (15 March 1954).

[41] PRO CAB 134/849, Chancellor of the Exchequer, *Reserve of Works* (14 December 1953).

[12] PRO CAB 134/848, *Studies of Measures to Combat a Recession in the US and to Strengthen the Economy*, Annex XII, Report by Working Group on Employment Policy (2 October 1953).

[43] PRO T230/229. *Preparations Against a US Recession*. On Robot see A. Cairncross, *Years of Recovery: British Economic Policy 1945–51* (London, 1985), Ch. 9.

[44] PRO T230/229, Economic Section, *Main Outlines of a Policy* (27

April 1953); Overseas Finance Section, *UK External Policy and US Recession* (27 April 1953).

[45] PRO T230/229, M. F. W. Hemming, *Comment* (30 April 1953).

[46] PRO T230/271, Chancellor of Exchequer, *Comments* (3 May 1953).

[47] PRO T229/591. *Implications of De-control* (21 August 1953).

[48] Ibid., *Interim Report on Import Cuts* (19 June 1953); compare ibid., Paymaster General, *The Power to Restrict Imports* (21 July 1953).

[49] E.g. PRO CAB 123/887, *UK Policy Towards Canada in the Event of US Recession* (25 September 1953).

[50] PRO CAB 134/852, *World Trade and Economic Conditions* (21 July 1954).

[51] Dow, *The Management of the British Economy*, 70.

9

Epilogue: The Decline and Fall of Employment Policy?

This chapter does not attempt a chronological account of employment policy over the last four decades. Rather, it sets out to explain in broad terms the decline of employment policy, from its period of development into the central policy objective (well established by the mid-1950s) to its explicit repudiation in the mid-1980s.

This change is well illustrated by the contents of two White Papers. The first is the 1944 Employment White Paper, already discussed at length. The second is the 1985 document, *Employment: The Challenge for the Nation*.[1] These papers are products of two different political contexts. The first is a broad statement of intent by a Coalition government, with no explicit legislative content. The second is much more a party political document, by a party on the defensive on the employment issue, and attempting to defend its political corner. Despite these different contexts, the White Papers both attempt to outline a general posture on employment policy, and it is this general posture that starkly summarizes the changes over the period.

The 1944 White Paper

The 1944 White Paper starts with the famous sentence, 'The Government accept as one of their primary aims and responsibilities the maintenance of a high and stable level of employment after the war.' In many ways, this sentence is the most revolutionary in the paper, for never before had such a responsibility been accepted by government.[2]

The second paragraph of the 1944 White Paper begins with the almost equally revolutionary statement that 'A country will not suffer from mass unemployment so long as the total demand for its goods and services is maintained at a high level.' Thus, from the beginning, the agenda is set in terms of a government responsibility to maintain demand. The paper then goes on to emphasize that external demand is as important as internal demand, and that the government is actively pursuing international policies to secure the external context for domestic full employment. The objectives of policy in this area are spelt out quite explicitly:

. . . to promote the beneficial exchange of goods and services between nations, to ensure reasonably stable rates of exchange, and to check the swings in world commodity prices which alternately inflate and destroy the incomes of the primary producers of foodstuffs and raw materials. It will also be necessary to arrange that countries which are faced with temporary difficulties in their balance of payments shall be able both to take exceptional measures to regulate their imports and to call on other nations, as good neighbours, to come to their help, so that their difficulties may be eased without recourse to measures which would permanently arrest the flow of international trade.[3]

These policy objectives in the international arena are coupled with an emphasis on the need for enhanced *manufacturing* efficiency, both for balance of payments reasons and to improve the standard of living. The government sees itself playing an active role in the encouragement of such efficiency.[4]

After a lengthy discussion of the transition period from war to peace, the White Paper 'turns aside from the main argument of the Paper to describe the measures which the Government will take to check the development of localised unemployment in particular industries and areas'.[5] The proposed solutions to this problem are seen as threefold: to influence the distribution of industry, to remove obstacles to labour mobility, and to enhance training for skills in expanding industries. It is in this context that the operation of the labour market is seen as important, and while the White Paper eschews large-scale labour transfers as a solution to localized unemployment, it suggests that full employment will provide the context for a loosening of restrictive practices which previously hindered interregional and inter-occupational mobility.[6]

The White Paper then goes on to outline the domestic conditions for a high and stable rate of employment. These again are seen as threefold:

(a) Total expenditure on goods and services must be prevented from falling to a level where general unemployment appears.

(b) The level of prices and wages must be kept reasonably stable.

(c) There must be a sufficient mobility of workers between occupations and localities.[7]

The maintenance of total expenditure has been dealt with in detail in previous chapters. On inflation, the emphasis in the 1944 White Paper is on the necessity for moderation in wage matters as a condition for success of policies of maintaining expenditure: 'If we are to operate with success on policy for maintaining a high and stable level of employment, it will be essential that employers and workers should exercise moderation in wage matters so that increased expenditure provided at the onset of a depression may go to increase the volume of employment.'[8]

The third element—labour mobility—is seen as crucial, because in its absence general expenditure measures would be ineffectual in reducing unemployment and would cause inflation.[9]

The White Paper then returns to the measures for maintaining general expenditure. This section includes the famous sentence that 'None of the main proposals contained in this Paper involves deliberate planning for a deficit in the National Budget in years of sub-normal trade activity.'[10] This is somewhat qualified by the acceptance that, even if in individual years the Budget is unbalanced, this does not imply any departure from balance over a longer period. The dangers of a long-run increase in the dead-weight debt, because of its effects on both the tax burden and on confidence, are also emphasized.[11]

The 1985 White Paper

The starting point for the 1985 White Paper is that unemployment in Britain reflects a failure of adaptation. 'Unemployment reflects our economy's failure to adjust to the circumstances and opportunities of today; to the changing pattern of consumer demand; to new competition from abroad; to innovation and technological development; and to world economic pressures.' The basic failure is of the labour market: 'The countries that have met this challenge successfully are those with efficient, competitive, innovative and responsive labour and goods markets. Improving the working of the labour market is particularly important.'[12]

The crucial role of government is to improve the working of the labour market, and thus to create a climate in which enterprise can flourish. Demand is explicitly *not* the problem, and the management of demand is concerned only with the control of inflation.[13]

Rather than a claim for responsibility, the 1985 White Paper emphasizes that the government can only 'set the framework for the nation's effort'. In support of this view, the 1944 White Paper is cited to the effect that employment cannot be created by government alone: 'But the success of the policy outlined in this Paper will ultimately depend on the understanding and support of the community as a whole and especially on the efforts of employers and workers in industry.'[14] However, as noted above, the 1944 document begins with a claim for responsibility, to which the cited passage is essentially a subordinate clause.

The 1985 White Paper goes on to argue that ever since the 1950s Britain's growth has been slower than its major competitors because of managerial inadequacies, union obstructionism, excessive government regulation, and the discouragement of the entrepreneur. These basic problems plus rapid inflation meant that Britain suffered much more from the oil shocks than other countries. The rise in unemployment in the 1970s and 1980s increased 'in large measure as the disguised unemployment of the earlier years, with overmanning rife, was forced into the open'.[15]

The current problem of reducing unemployment is seen as improving the response to unavoidable pressures from changes in the structure of employment, which reflects 'the biggest economic transformation since the first industrial revolution began two centuries ago'.[16] The government's policy response to this is threefold: first, the control of inflation to provide a stable framework for economic activity; second, deregulation and other improvements to the labour market; third, employment aid for particular groups.

Of these, it is the labour market that gains most attention. The 1944 White Paper is again cited in support of the importance of preventing inflation, via wage moderation, as a necessary step in maintaining employment.[17] As in the 1944 White Paper, no mechanisms are proposed to create this wage moderation; indeed, 'bureaucratic controls on pay' are specifically repudiated.[18]

The labour market needs to be improved via improvements in the quality of labour. This is to be pursued via improvements in school education, which has 'long underrated the central role of wealth-creating business in our national life', and by improvements in training, especially for young people.[19]

Labour market flexibility and costs are to be improved by changes in union laws, changes in taxes, and deregulation.

From 1944 to 1985

There are two immensely significant differences between the two White Papers. First, the 1944 document gives a central place to international measures seen as vital supports for domestic employment policy. There is a complex reformist strategy for international economic institutions aimed at providing a framework of exchange stability, free trade, and ample liquidity which will encourage trade and therefore employment stability and expansion.

The contrast with the 1985 posture is stark. Here, external events are treated as entirely exogenous, outside any control by policy. The only issue is the effectiveness of the response to these external changes. This is extraordinarily noticeable in the discussion of international competitiveness. Largely because of changes in the exchange rate, Britain suffered the worst recorded loss of competitiveness of any major country ever between 1977 and 1981.[20] Yet the discussion of competitiveness completely ignores this, in favour of focusing on labour costs.[21]

This stance is a logical corollary of saying that the sole function of macroeconomic policy is to control inflation—that there can be therefore no exchange rate or other external policy problem for macro-policy to deal with. The sole international question is one of competitiveness. Institutions are not at issue.

The second striking difference between the two papers is the movement from emphasis on demand to an emphasis on supply. This is clearly linked to external pressures, and hence the need for maintaining international demand. If this international orientation is given up, it is obviously very difficult to argue that a satisfactory policy can be pursued solely on the basis of regulating domestic demand. Thus, if the international economy is perceived to be beyond control, as simply an external force, it makes a certain sort of sense to focus on purely supply-side elements in the domestic economy.

The 1944 White Paper mentions almost all the supply-side policies that are given a central place in the 1985 White Paper. The difference is not that the earlier document ignores the supply side, especially the labour market, but that it argues that the necessary foundation of successful policies of this type is on a buoyant level of demand. And it accepts that government has a definite responsibility in this area, however much success may require changes in the behaviour of other groups.

Taken together, these two points bring out the extent to which the 1985 White Paper marks a retreat from government responsibility. Both internationally and domestically, the role of government is immensely narrowed, its perceived scope for successful action reduced almost to vanishing point. The sea change between 1944 and 1985 may then be summarized in one sentence as a perceived decline in the capacity of government to manage the economy; and it is this decline that the rest of this chapter tries to explain.

The International Economy and Economic Management

Interwoven with the development of employment policy in the early and mid-1940s was the reconstruction of the international economy. Largely at American initiative, though with significant British support, the attempt was made to build a system of world-wide multilateral trade supported by international financial institutions that would reduce pressures for protectionism. This drive for multilateralism was based on a very strong American perception that the Second World War had arisen as a direct result of the economic restrictionism—protection, exchange control—of the 1930s.[22] As in the case of nineteenth-century Britain, an open international market was seen as felicitously combining export expansion for the home economy and the conditions for international peace.

Much of this effort to reconstruct the world economy was naive in conception, and the institutions intended to underpin it either were never born (the International Trade Organization) or were strangled soon after birth (the Bretton Woods Agreement). Almost from the beginning, it was apparent that these grand designs took little account of the scale of the postwar economic pre-eminence of the United States, or of the scale of the reconstruction effort that was to be pursued in Western Europe. Together, these made the transition to a multilateral world much more painful and prolonged than any one envisaged in the period of postwar planning during the war.[23]

Nevertheless, by the late 1950s the Western Hemisphere, at least, was moving towards a world based on the principles, if not the institutions, that had been espoused in the mid-1940s.[24] Despite the creation of the EEC and the continuation of some British Commonwealth preferences, the general trend was towards much freer trade helped by the General Agreement on Trade and Tariffs (GATT), the 'temporary' replacement for the ITO. Exchange rate stability was well established, and from the late 1950s exchange controls were largely abolished; the International Monetary Fund (IMF) came to occupy something like the role originally envisaged at Bretton Woods.

It is important to emphasize that the American emphasis was on multilateral trade as the centrepiece of the grand design.[25] International financial agreements were seen as a buttress to this trade system. There was little enthusiasm for encouraging international capital flows. While the potential role of short-term funds in correcting short-term balance of payments difficulties was accepted, there was a very strong awareness of the role of 'hot-money' flows in destabilizing the interwar economy.[26] Equally, while there was a growing concern with providing long-term capital to underdeveloped countries from the early 1940s, such flows between the Western countries were not an issue in the original grand design.

There were disagreements about how far exchange rate changes were desirable to avoid deflationary responses to external disequilibrium, and in the event a system of fairly fixed rates became established.[27] This had the very important effect of encouraging private international capital flows on a scale that was eventually to match pre-1914 levels (for long-term capital) and far surpass anything in the past (for short-term flows).

These flows were not just the unintended consequence of the relative fixity of exchange rates. They reflected also the growth in world trade, coupled with that of multinational enterprises. Overall, the Western world since the 1950s has been characterized by a growth of economic interdependence to an unparalleled extent.[28]

This interdependence has undoubtedly made the constraints on national economic management much tighter. Of the three elements noted above, probably the growth of multinational enterprises is the least important. While this development has facilitated the transfer of production facilities between nations, it is far from clear that this is an independent constraint on national policy. In other words, if a country has a relatively successful manufacturing capacity, then a high level of multinational production may prove to be no problem. This would seem to be the case in, for example, Sweden.[29]

Dependence on foreign markets and imports from abroad undoubtedly increases the constraints on economic management. As will be returned to below, inflation in particular has become increasingly an international phenomenon, partly because of the dependence of Western countries on imports for many of their raw materials and fuels. But the most direct and compelling constraint on domestic and economic management would seem to be the scale and volatility of international capital flows.

This constraint operates at different levels. It means that governments pursuing domestic policies that are unpopular with international financial opinion may find it difficult to borrow abroad. Something of this sort occurred, of course, in 1931 when the Labour government fell over the conditions asked for by potential lenders to Britain. The same process was at work in 1976, only this time the result was an acceptance of conditions asked for by the lender—the IMF.[30]

More broadly, in the 1970s and 1980s there has emerged a general asymmetry in the international financial response to domestic policies. Reflationary domestic policies (Britain 1975/6, France 1981/2) tend to be undermined by capital flight, exchange depreciation, and increasing inflation.[31] Policy reversal then becomes very difficult to avoid. Conversely, deflationary domestic policies are reinforced by capital inflows, exchange appreciation, and downward pressure on inflation.[32]

This pattern is very much a new one in the 1970s and 1980s. While the growing financial interdependence that underlies it was developing in the 1950s and 1960s, other conditions in that period prevented it from manifesting itself as a major constraint on domestic policies. Above all, this was because of the underlying buoyancy of the North American and European economies, which meant that high employment and economic expansion could be obtained *without* radical domestic policies.

The Keynesian revolution in economic theory may have helped

to create a climate in which long-run deflationary responses to balance of payments difficulties or inflation became less conceivable. It may have also helped investors to believe that governments would maintain high demand, and hence may have encouraged investment. The fact that no major threat to employment emerged 'does not necessarily invalidate the notion that business confidence and private spending were supported by the belief that, if it had, the means were at hand to counteract it'.[33]

However, this does seem a rather desperate attempt to save the phenomenon of a 'Keynesian revolution' in economic policy. Rather, the basic point would seem to be that expansion was due to 'autonomous' growth of investment, buttressed by trade growth.[34] In this climate, governments could commit themselves, with considerable differences of method and emphasis, to expansionary policies because 'that expansion proved possible without posing political and economic problems which were too great to resolve'.[35]

The problem of the 1970s and 1980s was precisely that such expansion and commitment to full employment came to pose larger and larger problems in these decades. The demise of employment policy was *one way* in which these problems could be 'resolved' politically.

The central political difficulty that grew sharply in significance in the 1970s was inflation. Undoubtedly, the high levels of employment of the postwar period created an inflationary bias in most countries, both by the direct effect on wage bargaining and by helping to create a climate of expectations about future prosperity. But this bias was contained in the 1950s and 1960s, aided perhaps by unprecedentedly sustained real-wage growth and also, perhaps, by the fixed exchange rate regime.[36]

Internationally, too, the forces for inflation were muted in the 1950s and early 1960s. There was, of course, a very substantial rise in the price of primary commodities during the Korean war period. But, unlike the similar episode in the early 1970s, this was quickly reversed.[37] Only from the late 1960s did the expansionary effects of US military expansion in Vietnam ripple through the world economy. This expansion also added new impetus to the deterioration of the US international economic position, which, coupled with inflation, was to end this period of a fixed exchange rate system at the beginning of the 1970s.

In the early 1970s, and again at the end of that decade, international inflations were unleashed by the concurrent expansion of most Western economies, with sharp effects on the price of internationally traded primary commodities, exacerbated by the efforts of OPEC. Inflation levels in individual countries were not completely at the mercy of these international price movements. Indeed, the variance of inflation rates grew in the 1970s—but around a similar trend in almost all countries.[38] A similar point could be made about the slowdown in inflation in the early and mid-1980s—it was predominantly, though not exclusively, a consequence of the reversal of the previous rise in commodity prices.[39]

Although we may judge that the scope for individual governments to control the rate of inflation under current conditions is relatively limited, this did not reduce the political pressure of national governments to do something about the problem.

Dealing with Inflation in Britain

From the earliest discussions of employment policy, it was recognized by both economists and politicians that this would impart an inflationary bias to the economy.[40] Equally, there was widespread acceptance of the difficulties of violating the conditions of 'free collective bargaining' in British conditions.[41] Hence the stage was set for the characteristic British approach of periodic crisis-borne incomes policies. These lasted only as long as the immediate crisis. More fundamentally, they were never linked to any strategy for the labour market which might be able to alter the long-run possibilities of coping with external shocks to the economy. This failure is most marked for the left, for in other European countries it was notably forces of the left that were able to formulate such strategies and thus ride the storms of the world economy much more successfully. This is not to suggest that the success of national economic management in for example Sweden or Austria was simply a consequence of wages policies; nor is it to suggest that such policies could be easily transplanted into British institutional conditions. Rather, the point here is to register certain characteristics of the domestic economy policy regime in Britain which opened the way for the demise of employment policy. The failure of the left to offer a coherent and sustainable alternative to the right in responding to inflationary pressures in the labour market was, it is suggested, central to that demise. This *political* point obtains, even if it is the case that much inflationary pressure arose outside the labour market.

The Social Contract of the 1970s was less of a short-term crisis measure than the incomes policies that preceded it. But except for some egalitarian aspects, it was not based on any fundamental reappraisal of the labour market by the left, or on a reappraisal of the role of the trade unions that such an approach would necessitate.[42] The triumph of the Conservatives in 1979 was not borne of a lack of concern with unemployment, or of a widespread commitment to the anti-welfare rhetoric of 'Thatcherism': rather, it resulted from the failure of the previous government to sustain its efforts at economic management (successful on the inflation and employment fronts in 1977 and 1978) because of a failure to carry the trade unions with the policy.[43]

The Basis of the 1985 White Paper

The 1985 employment document represents a particular way of coming to terms with the constraints on domestic economic management. Broadly, this coming to terms involves accepting the constraints as largely beyond any control, and treating any judgement on domestic policy by international capital as manifesting the highest rationality. Hence the simple acceptance of the extraordinary appreciation of the pound in 1979 and 1980: in any event, it would have been a difficult episode to manage, but the government adopted the stance of positively welcoming this endorsement of their deflationary stance. This appreciation of the exchange rate was the mechanism whereby an unprecedented rise in unemployment occurred—a doubling in the space of eighteen months.

Almost incredibly, the 1985 White Paper does not refer to this episode, or indeed to the general problem of the exchange rate in relation to unemployment. Competitiveness is discussed solely in relation to labour costs and productivity, elements that were completely overwhelmed in 1979–80 by exchange rate changes.[44]

In the 1985 document external events are treated as data, the only issue being how well the British economy adapts to them.

There is no mention of international economic institutions except for the endorsement of free convertibility of the pound. This of course is in striking contrast with the 1944 White Paper, which contained a great deal on international economic conditions, and was accompanied by enormous efforts to shape a world economy compatible with a domestic full-employment policy.

To emphasize structural changes in the world economy as the most fundamental element in the shift from the White Paper of 1944 to that of 1985 is not to suggest these changes as the only condition. Clearly, the 1944 White Paper was also conditioned by the broad shift in public opinion towards the left, which in this context meant a very strong belief that the economy, by purposive management, could and should be improved in its performance over that in the 1930s. This shift in opinion was accompanied at the more intellectual level by a belief in a tehnocratic and rationalistic approach to economic and social policy. By the mid-1970s, the parallel rise in unemployment and inflation had served to undermine faith in the capacity of governments to deliver policy objectives. This was reflected at the intellectual level in a resurgence of faith in *laissez-faire* solutions to economic and social policy.

This point should not be overdrawn. The Conservative victory in 1979, the immediate political condition for the 1985 White Paper stance, was not the consequence of a massive shift in public opinion, but rather, as suggested above, reflected concern at the economic management failures of the previous Labour government, coupled with the interwoven and longer-running issue of the role of the trade unions. Equally, the role of the intelligentsia and shifts in their opinions should not be exaggerated in their consequences. In particular, probably too much weight has been given to shifts in academic economics towards monetarism and *laissez-faire*. While this undoubtedly played its part in the demise in the full-employment commitment, the commitment was rendered much more problematic even in countries where monetarist ideology was never a force.

Monetarism, it might be argued, can be seen in part as a reaction to the utopianism of some Keynesian advocacy, which was legitimized in the 1950s and 1960s by the delivery of expansion and full employment without political pain, but was undermined by the events of the 1970s and 1980s. This is not to argue that monetarism was 'right', but that the space for its growth was created by the Keynesian failure to cope with the changes in the world economy, and their impact on the constraints and possibilities of national economic management.

This utopianism was most apparent in the attitude to budget deficits, which were seen as essentially a technical economic device. If they could be justified on grounds of economic theory, then that was the end of the problem. Such a rationalistic view was tenable in the 1950s and 1960s, when fiscal policy involved (in Britain) current account surpluses on the Budget and a small public sector borrowing requirement (PSBR).[45] But in circumstances such as the mid-1970s, when a Keynesian stance implied a much more radical fiscal policy, all these problems emerged. In many respects, they were those problems addressed in the period before the 1944 White Paper (see Chapter 3 above): issues of confidence, of the response of private investors to public policy, of the capacity of politicians to evade 'responsible' financial policy via budget deficits.[46] Keynesianism cannot be reduced to budgetary policy; but the failure here to come to grips with the old questions ceded the ground to those who wanted to argue that such policies were undesirable as well as infeasible.

The 1985 White Paper also had directly political components. Its focus on the labour market fitted with a longstanding Conservative hostility to trade unions. Its macreconomic focus on the PSBR linked with a longstanding desire to reduce the size of the public sector (though less easily with a concern to reduce taxes). The package was not without its problems or its potential contradictions. Above all, it is far from clear that the labour market was fundamentally changed by the policies of the early 1980s, given the behaviour of real wages in the mid-1980s. But overall, it was a complete and relatively coherent rejection of the world view of the 1944 White Paper.

Conclusions

This chapter has, in a broad-brush way, attempted to explain why the 1985 White Paper is so strikingly different from its predecessor of 1944. The answer suggested is not intended to be of a monocausal kind. It is a question of emphasis. Here the emphasis has been on the changed conditions of the international economy, the limits these have imposed on national economic management, and the ways in which they have been come to terms with.

The rejection of demand management and macroeconomic policy by the 1985 White Paper is one way in which these constraints can be lived with. It resolves the conflict between employment policy and international constraints by abolishing employment policy. Plainly, this is not the only possible resolution. An alternative path would involve a much more activist policy, both to reduce the constraints imposed by international conditions by changes in international economic policy and institutions, and domestically to come to terms with those constraints in a realistic way. This is a formidable task, above all calling for an exercise of *political* intelligence on the left which, on past standards, it appears rather optimistic to hope for.

Notes

[1] Cmnd. 9474. The 1944 White Paper is referred to as Cmd. 6527.

[2] It is perhaps worth emphasizing that such a commitment, with varying degrees of qualification, was made in most advanced capitalist countries around this time.

[3] Cmd. 6527, para. 3.

[4] Ibid., para. 6.

[5] Ibid., para. 20.

[6] Ibid., para. 20–36.

[7] Ibid., para. 39.

[8] Ibid., para. 49.

[9] Ibid., para. 56.

[10] Ibid., para. 74.

[11] Ibid., paras. 76–9.

[12] Cmnd. 9474, para. 1.3.

[13] Ibid., paras. 1,4, 2.3, 5.1.

[14] Ibid., para. 1.5, citing Cmd. 6527, Foreword.

[15] Ibid., para. 2.7.

[16] Ibid., para. 4.1.

[17] Ibid., para. 5.3, citing Cmd. 6527, para. 49.

[18] Ibid., para. 4.4; also para. 5.5

[19] Ibid., para. 6.2

[20] See e.g. R. Backhouse, *Macroeconomics and the British Economy* (Oxford, 1983), 58–62, 267–8.

[21] E.g. para. 7.7. This focus on labour is part of a national obsession, exemplified by the Edwarde's period at British Leyland, the disastrous

consequences of which are described in K. Williams, J. Williams, and C. Haslam, *The Breakdown of Austin Rover* (London, 1987).

[22] R. N. Gardner, *Sterling–Dollar Diplomacy* (Oxford, 1956), Ch. 1.

[23] Ibid. See also A. S. Milward, *The Reconstruction of Western Europe 1945–51* (London, 1985). Milward emphasizes the abortive nature of the Bretton Woods agreement given the political imperative for economic expansion in Western Europe, e.g. 50–4, 463–4.

[24] Milward, *Reconstruction*, 44. Note that Gardner, writing in the mid-1950s, had to record at that time that the American grand design had been a failure: Gardner, *Sterling–Dollar Diplomacy*, Ch. 18.

[25] Not 'free trade' but non-discrimination in trade was the central thrust: Gardner, *Sterling–Dollar Diplomacy*, 13.

[26] League of Nations, *International Currency Experience* (Geneva, 1944); also R. N. Cooper, *The Economics of Interdependence* (New York, 1968), 27–8.

[27] The original US 'White' plan had envisaged little room for exchange rate changes, because of fears of competitive devaluations as occurred in the 1930s. This was later amended because of British fears that such fixity would force unemployment on countries with weak balance of payments: Gardner, *Sterling–Dollar Diplomacy*, 14–16.

[28] M. Stewart, *Controlling the Economic Future: Policy Dilemmas in a Shrinking World* (Brighton, 1983), Ch. 3.

[29] B. Swedenborg, 'Sweden', in J. H. Dunning (ed.), *Multinational Enterprise, Economic Structure, and International Competitiveness* (New York, 1985).

[30] J. Tomlinson, *British Macroeconomic Policy since 1940* (London, 1985), 126–36, 155–61.

[31] This did not seem to apply to the United States in the early 1980s. Partly this was perhaps because the reflationary fiscal policy was coupled with other highly conservative policies. Also, with other Western countries pursuing conservative fiscal policies, the United States alone provided the public debt to soak up the contractual savings of the West.

[32] Stewart, *Controlling the Economic Future*, Chs. 5, 6.

[35] J. Bispham and A. Boltho, 'Demand Management', in A. Boltho (ed.), *The European Economy: Growth and Crisis* (Oxford, 1982), 320.

[34] R. C. O. Matthews: 'Why Has Britain Had Full Employment Since the War?', *Economic Journal*, 78 (1968), 555–69.

[35] Milward, *Reconstruction*, 478.

[36] Fixed exchange rates may have helped the containing of inflation, as fast inflation showed up rapidly in balance of payments difficulties, which were politically easy to use to pursue (short-term) anti-inflationary policies.

[37] C. Allsopp, 'Inflation', in Boltho, *The European Economy*, 80–1.

[38] Ibid., Table 3.2, 79.

[39] W. Beckerman and T. Jenkinson, 'What Stopped Inflation: Unemployment or Commodity Prices?' *Economic Journal*, 96 (1986), 39–54.

[40] Jones, *Wages and Employment Policy, 1936–86* (London, 1987).

[41] Some discussion of the strength of this notion in Britain is given in J. Tomlinson, *Monetarism: Is There an Alternative?* (Oxford, 1986), Ch. 3.

[42] Ibid.

[43] D. E. Butler and D. Kavanagh, *The British General Election of 1979* (Oxford, 1980).

[44] Cmnd. 9474, paras. 7.7–7.10.

[45] Tomlinson, *British Macroeconomic Policy*.

[46] M. Holmes, *The Labour Government 1974–79: Political Issues and Economic Reality*, (London, 1986).

Bibliography

Public Records

Public Record Office, Cabinet Papers, CAB 21, Prime Minister's Briefs.
Public Record Office, Cabinet Papers, CAB 66, Cabinet Committees.
Public Record Office, Cabinet Papers, CAB 124, Lord President of the Council's Files.
Public Record Office, Cabinet Papers, CAB 128, Cabinet Conclusions.
Public Record Office, Cabinet Papers, CAB 129, Cabinet Memoranda.
Public Record Office, Cabinet Papers, CAB 130, Cabinet Committees.
Public Record Office, Cabinet Papers, CAB 132, Lord President's Committee and Sub-committees.
Public Record Office, Cabinet Papers, CAB 134, Cabinet Committees.
Public Record Office, Prime Minister's Office, PREM 4 (1939–45).
Public Record Office, Prime Minister's Office, PREM 8 (1945–51).
Public Record Office, Treasury, T160, Finance.
Public Record Office, Treasury, T161, Supply.
Public Record Office, Treasury, T171, Budget and Finance Bill.
Public Record Office, Treasury, T228, Trade and Industry.
Public Record Office, Treasury, T229, Central Economic Planning Staff.
Public Record Office, Treasury, T230, Economic Section.
Public Record Office, Treasury, T247, Keynes Papers.

Official Reports and Papers

1st Interim Report of the Committee on Currency and Foreign Exchange After the War, Cd. 9182 (1918).
Memoranda on Certain Proposals Relating to Unemployment, Cmd. 3331 (1929).
Social Insurance and Allied Services, Cmd. 6404 (1942).
Employment Policy, Cmd. 6527 (1944).
Report of the Committee on the Working of the Monetary System, Cmnd. 827 (1959).
Employment: The Challenge for the Nation, Cmnd. 9474 (1985).

Unpublished Work

Meade, J. E., 'Diaries' (London School of Economics).
Macleod, R. J., 'The Development of Full Employment Policy, 1938–45', D. Phil. dissertation (Oxford, 1978).

Books

Addison, P., *The Road to 1945* (London, 1977).
Aldcroft, D. H., *The British Economy, i. The Years of Turmoil 1920–51* (Brighton, 1986).
Allsop, C., 'Inflation', in A. Boltho (ed.), *The European Economy: Growth and Crisis* (Oxford, 1982).
Arndt, H. W., *Economic Consequences of the Nineteen Thirties* (New York, 1944).
Backhouse, R., *Macroeconomics and the British Economy* (Oxford, 1983).
Balogh, T., 'The International Aspect', in G. D. N. Worswick and P. H. Ady (eds.), *The British Economy 1945–50* (Oxford, 1952).
Barnett, C., *The Audit of War* (London, 1986).
Beveridge, W. H., *Full Employment in a Free Society* (London, 1944).
Bispham, J., and Boltho, A., 'Demand Management', in A. Boltho (ed.), *The European Economy: Growth and Crisis* (Oxford, 1982).
Booth, A., and Pack, M., *Employment, Capital, and Economic Policy* (Oxford, 1985).
Briggs, A., 'The Political Scene', in S. Nowell-Smith (ed.), *Edwardian England 1900–1914* (Oxford, 1964).
Brown, A. J., *The Great Inflation 1939–51* (Oxford, 1953).
Brown, K. D., *Labour and Unemployment* (Newton Abbot, Devon, 1971).
Buchanan, J., and Wagner, R., *Democracy in Deficit: The Political Legacy of Lord Keynes* (London, 1977).
Butler, D. E., *The British General Election of 1951* (London, 1952).
—— and Kavanagh, D., *The British General Election of 1979* (Oxford, 1980).
Butler, R. A., *The Art of the Possible* (London, 1971).
Cairncross, A., *Years of Recovery: British Economic Policy 1945–51* (London, 1985).
Chester, D. N., 'The Central Machinery for Economic Planning', in D. N. Chester (ed.), *Lessons of the British War Economy* (London, 1951).
—— *The Nationalisation of British Industry 1945–51* (London, 1975).

Clarke, R. W. B., *Ango-American Collaboration in War and Peace 1942–9* (Oxford, 1982).

Cooper, R. N., *The Economics of Interdependence* (New York, 1968).

Cowling, M., *The Impact of Labour 1920–24* (Cambridge, 1971).

—— *The Impact of Hitler* (Cambridge, 1975).

Craig, F. W. S., *British General Election Manifestos 1918–66* (Chichester, 1970).

Cutler, A., Williams, K., and Williams, J., *Keynes, Beveridge, and Beyond* (London, 1986).

Dalton, H., *Principles of Public Finance* (4th edn.; London, 1954).

Davison, R. C., *The Unemployed: Old Policies and New* (London, 1929).

Devons, E., 'The Problem of Co-ordination in Aircraft Production', in D. N. Chester (ed.), *Lessons of the British War Economy* (London, 1951).

Dow, J. C. R., *The Management of the British Economy 1945–60*, (Cambridge, 1965).

Feinstein, C. H., *National Income, Expenditure and Output of the United Kingdom 1855–1965* (Cambridge, 1972).

Ford, A., *The Gold Standard 1880–1914: Britain and Argentina* (Oxford, 1962).

Gardener, R. N., *Sterling–Dollar Diplomacy* (Oxford, 1956).

Garratty, J., *Unemployment in History: Economic Thought and Public Policy*, New York, 1978).

Hancock, W. K., and Gowing, M., *British War Economy* (London, 1949).

Harris, J., *Unemployment and Politics: A Study in English Social Policy, 1886–1914* (Oxford, 1972).

—— *William Beveridge: A Biography* (Oxford, 1977).

Holmes, M., *The Labour Government 1974–79: Political Issues and Economic Reality* (London, 1986).

Howson, S., *Domestic Monetary Management in Britain 1919–38* (Cambridge, 1975).

—— and Winch, D., *The Economic Advisory Council 1930–39: A Study in Economic Advice during Depression* (Cambridge, 1977).

Hutchinson, T. W., *Economics and Economic Policy in Britain 1946–66* (New York, 1970).

Jay, D., *Change and Fortune: A Political Record* (London, 1980).

Jewkes, J., 'Second Thoughts on the British White Paper on Employment Policy', in National Bureau of Economic Research, *Economic Research and the Development of Economic Science and Public Policy* (New York, 1946).

—— *Ordeal by Planning* (London, 1948).

—— *A Return to Free Market Economics?* (London, 1978).

Jones, R., Wages and Employment Policy, 1936–86 (London, 1987).
Kennedy, C., 'Monetary Policy', in G. D. N. Worswick and P. H. Ady (eds.), *The British Economy 1945–50* (Oxford, 1952).
Kennedy, C., 'Monetary Policy', in G. D. N. Warwick and P. H. Ady (eds.), *The British Economy in the 1950s* (Oxford, 1962).
Keynes, J. M. (ed.), *Alfred Marshall's Official Papers* (London, 1926).
—— *The Collected Writings of J. M. Keynes*, iv. *A Tract on Monetary Reform* (London, 1971).
—— *The Collected Writings of J. M. Keynes*, vii. *The General Theory of Employment, Interest and Money* (London, 1973).
—— *The Collected Writings of J. M. Keynes*, xx. *Activities 1929–31: Rethinking Employment and Unemployment Policies* (London, 1981).
—— *The Collected Writings of J. M. Keynes*, xxii. *Activities 1939–45; Internal War Finance* (London, 1978).
—— *The Collected Writings of J. M. Keynes*, xxv. *Activities 1940–44; Shaping the Post War World: The Clearing Union* (London, 1980).
—— *The Collected Writings of J. M. Keynes*, xxvii. *Activities 1940–46; Shaping the Post War World: Employment and Commodities* (London, 1980).
League of Nations, *International Currency Experience* (Geneva, 1944).
Lerner, A. *Economics of Control* (London, 1944).
Little, I. M. D., 'Fiscal Policy', in G. D. N. Worswick and P. H. Ady (eds.), *The British Economy 1945–50* (Oxford, 1952).
Macdougall, D., 'The Prime Minister's Statistical Section', in D. N. Chester (ed.), *Lessons of the British War Economy* (London, 1951).
Maddison, A., *Economic Growth in the West* (London, 1964).
Meade, J., *An Introduction to Economic Analysis and Policy* (2nd edn; London, 1937).
Middlemas, K., 'Edwardian Socialism', in D. Read (ed.), *Edwardian England*, (London, 1982).
Middleton, R., *Towards the Managed Economy: Keynes, the Treasury, and Economic Policy in the 1930s* (London, 1985).
Milward, A. S., *War, Economy and Society 1939–45* (London, 1977).
—— *The Reconstruction of Western Europe 1945–51* (London, 1985).
Morgan, K. O., *Consensus and Disunity: The Lloyd George Coalition Government 1918–22* (Oxford, 1979).
Morley, J., *The Life of Richard Cobden* (London, 1905).
Pigou, A. C., *Riddle of the Tariff* (London, 1904).
—— *British Economic History 1914–1925* (London, 1947).
Pilgrim Trust, *Men Without Work* (Cambridge, 1938).
Pimlott, B., *Hugh Dalton* (London, 1985).
Pollard, S. (ed.), *The Gold Standard and Employment Policies Between the Wars* (London, 1970).
Sayers, R. S., *British Financial Policy 1939–45* (London, 1956).

—— 1941—The First Keynesian Budget', in C. H. Feinstein (ed.), *The Managed Economy* (Oxford, 1983).

Scott, M. F., 'The Balance of Payments Crises', in G. D. N. Worswick and P. H. Ady (eds.), *The British Economy 1945–50* (Oxford, 1952).

Semmel, B., *Imperialism and Social Reform* (London, 1960).

Skidelsky, R., *Politicians and the Slump* (Harmondsworth, Middx., 1970).

Skocpol, T., and Weir, M., 'State Structures and the Possibilities for "Keynesian" Responses to the Great Depression in Sweden, Britain, and the United States', in P. Evans, D. Rueschmayer, and T. Skocpol (eds.), *Bringing the State Back In* (Cambridge, 1985).

Snowden, P., *Labour and National Finance* (London, 1920).

Stewart, M., *Keynes and After* (Harmondsworth, Middx.; 1967).

—— *Controlling the Economic Future: Policy Dilemmas in a Shrinking World* (Brighton, 1983).

Strange, S., *Sterling and British Policy* (London, 1971).

Streeton, P., 'Commercial Policy', in G. D. N. Worswick and P. H. Ady (eds.), *The British Economy 1945–50* (Oxford, 1952).

Swedenborg, B., 'Sweden', in J. H. Dunning (ed.), *Multinational Enterprise, Economic Structure, and International Competitiveness* (New York, 1985).

Thirlwall, A. P. (ed.), *Keynes as a Policy Adviser* (London, 1982).

Thomas, T. J., 'Aggregate Demand in the United Kingdom, 1918–45', in R. Floud and D. McCloskey (eds.), *The Economic History of Britain since 1750*, ii. *1860 to the 1970s* (Cambridge, 1981).

Thompson, G., *The Conservatives' Economic Policy* (London, 1986).

Tomlinson, J., *Problems of British Economic Policy, 1870–1945* (London, 1981).

—— *British Macroeconomic Policy since 1940* (London, 1985).

—— *Monetarism: Is There an Alternative?* (Oxford, 1986).

Williams, K., Williams, J., and Haslam, C., *The Breakdown of Austin Rover* (London, 1987).

Wilson, T., 'Policy in War and Peace: The Recommendations of J. M. Keynes', in A. P. Thirlwall (ed.), *Keynes as a Policy Adviser* (London, 1982).

Winch, D., *Economics and Policy: A Historical Survey* (rev. edn.; London, 1972).

Worswick, G. D. N., 'The British Economy 1950–59', in G. D. N. Worswick and P. H. Ady (eds.), *The British Economy in the 1950s* (Oxford, 1962).

Articles

Beckerman, W., and Jenkinson, T., 'What Stopped Inflation: Unemployment or Commodity Prices?' *Economic Journal*, 96 (1986), 39–54.

Booth, A., 'The "Keynesian Revolution" in Economic Policy-Making', *Economic History Review*, 36 (1983), 103–23.

—— 'Defining a "Keynesian Revolution" ', *Economic History Review*, 37 (1984), 263–8.

—— 'The "Keynesian Revolution" in Economic Policy-Making: A Reply', *Economic History Review*, 38 (1985), 101–6.

—— 'Economists and Points Rationing in the Second World War', *Journal of European Economic History*, 14 (1985), 297–317.

—— 'Simple Keynesianism and Whitehall', *Economy and Society*, 15 (1986), 1–22.

——, and Glynn, S., 'Unemployment in Interwar Britain: A Case for Relearning the Lessons of the 1930s?', *Economic History Review*, 36 (1983), 329–48.

—— 'Building Conterfactual Pyramids', *Economic History Review*, 38 (1985), 89–94.

Cairncross, A., 'An Early Think Tank: Origins of the Economic Section', *Three Banks Review*, 144 (1984), 50–9.

Coats, A. W., 'Political Economy and the Tariff Reform Campaign of 1903', *Journal of Law and Economics*, 11 (1968), 181–229.

Garside, W. R., 'The Failure of the "Radical Alternative": Public Works, Deficit Finance and British Interwar Unemployment', *Journal of European Economic History*, 14 (1985), 537–55.

——, and Hatton, T. J., 'Keynesian Policy and British Unemployment in the 1930s', *Economic History Review*, 38 (1985), 38–8.

Glynn, S., and Howells, P., 'Unemployment in the 1930s: the Keynesian "Solution" Reconsidered', *Australian Economic History Review*, 20 (1980), 28–45.

Hirst, P., 'The Necessity of Theory', *Economy and Society*, 8 (1979), 417–45.

Jones, M. E. F., 'Regional Employment Multipliers, Regional Policy, and Structural Change in Interwar Britain', *Explorations in Economic History*, 22 (1985), 417–39.

Matthews, R. C. O., 'Why Has Britain Had Full Employment Since the War?' *Economic Journal*, 78 (1968), 555–69.

Middleton, R., 'The Treasury in the 1930s: Political and Administrative Constraints to Acceptance of the "New" Economics', *Oxford Economic Papers*, 34 (1982), 48–77.

Peden, G., 'Keynes, the Treasury and Unemployment in the Later Nineteen Thirties', *Oxford Economic Papers*, 32 (1980), 1–18.

—— 'Sir Richard Hopkins and the "Keynesian Revolution" in Employment Policy 1929–45', *Economic History Review*, 36 (1983), 281–96.

Rollings, N., 'The "Keynesian Revolution" in Economic Policy-making: A Comment', *Economic History Review*, 38 (1985), 95–100.

Smith, K., 'Why Was There Never a "Keynesian Revolution" in

Economic Policy?: A Comment', *Economy and Society*, 11 (1982), 223–8.

Thomas, M., 'Rearmament and Economic Recovery in the late 1930s', *Economic History Review*, 36 (1983), 552–79.

Tomlinson, J., 'Why Was There Never a "Keynesian Revolution" in Economic Policy?', *Economy and Society*, 10 (1981), 72–87.

—— 'Women as Anomalies: The Anomalies Regulations Act of 1931, Its Background and Implications', *Public Administration*, 62 (1984), 423–37.

Index